KEEPING A FAMILY COW

KEEPING A FAMILY COW

Revised and Updated Edition

The Complete Guide for Home-Scale, Holistic Dairy Producers

Joann S. Grohman

Foreword by Jack Lazor

Chelsea Green Publishing
White River Junction, Vermont

First published as *The Cow Economy* in 1975 with Merril Grohman (Coburn Farm Press) and as *Keeping a Family Cow*, by Joann S. Grohman, in 1981 (Charles Scribner's Sons), in 1984 (Ballantine Books), and in 1999, 2001, 2003, 2007, and 2008 (Coburn Press, 434 Carthage Road, Carthage, ME 04224).

Project Manager: Hillary Gregory
Editor: Makenna Goodman
Copy Editor: Nancy Ringer
Proofreader: Eileen M. Clawson
Indexer: Shana Milkie
Designer: Melissa Jacobson

Printed in the United States of America.
First printing September, 2013.
10 9 8 7 6 5 19 20 21 22

Library of Congress Cataloging-in-Publication Data
Grohman, Joann S.
Keeping a family cow / by Joann S. Grohman ; foreword by Jack Lazor. – Rev. and updated ed.
p. cm.
Includes bibliographical references and index.
ISBN 978-1-60358-478-4 (pbk.) – ISBN 978-1-60358-479-1 (ebook)
1. Dairy farming. 2. Cows. 3. Dairying. I. Title.

SF239.G76 2013
636.2'142 – dc23

2013022230

Chelsea Green Publishing
85 North Main Street, Suite 120
White River Junction, VT 05001
(802) 295-6300
www.chelseagreen.com

Contents

Foreword

In May of 1975, my wife, Anne, and I rented a little farm in Irasburg, Vermont. Within a week, we had purchased a family cow named "Pet." It was all part of our master plan – we were going back to the land in search of a meaningful life. We wanted to provide for ourselves independent of a world that we saw heading in a different direction.

We thought we were ready for the task at hand. I had read lots of books and had studied the history of agriculture while at college. Anne had taken several animal science courses at UW Madison. We had both done stints at dairy farms and had worked on the historical farm at Old Sturbridge Village. We were so wet behind the ears. What we lacked in experience, we made up with idealism and ambition.

It was right about this time that we came across a little book called *The Cow Economy* by Merril and Joann Grohman – a book about the very lifestyle we wanted to lead. Everything we needed to know about keeping a family cow was in this little book. Feeding, breeding, and the general principles of animal husbandry were all discussed right down to last nuanced detail. With help from our friends and neighbors and the advice contained in this little book, we made our way into the world of dairy farming.

Our cow Pet had Sundance. Two years later Sundance had her own heifer calf, and before we knew it, we were raising a group of heifer calves on the surplus milk. By 1979, we were milking six cows, making dairy products on our kitchen stove, and delivering them to our local community. In 1984, we obtained a milk handler's license from the State of Vermont. Our dairy herd had increased to twelve Jersey cows by this time. Somehow, as the years had passed, we had transitioned from being homesteader "back-to-the-land" family cow types into commercial dairy farmers. Fewer families were keeping cows at this time, and it provided us with a wonderful opportunity to start a small dairy. The demand for wholesome farm-produced dairy products was there, and we stepped up to the plate.

Forty years have gone by since we bought our first cow, and we have seen many changes during this time. We have sharpened our farming skills and improved our herd and our land. We have developed an innate understanding of the concepts of organic farming. As our knowledge base has increased, we have shared our experiences with others in the field of agriculture. A burgeoning local foods movement has developed with a general public hungry for meaning in the food they eat. Micro-dairies, like the one we started forty years ago, are springing up everywhere. The family cow is coming back.

Considering this new trend in sustainability and local food consumption, it is only fitting that Joann Grohman's book *The Cow Economy* is being reissued to provide inspiration and information to a new group of latter-day homesteaders who are now just going back to the land. This is no ordinary text. Republished as *Keeping a Family Cow*, this book combines food philosophy with a practicum of knowledge and experience that Ms. Grohman has acquired in her eighty-five years in and around Jersey cows. Joann's book is a field manual for both the experienced and inexperienced alike. Everything you ever wanted to know about cows and more is covered in this volume. The basics of animal husbandry are discussed as they were in the original edition, along with a lot of newer information about organic practices and current concerns in the world of milk. If you want to know about haymaking or dairy-product production, it's all in here.

Anne and I recently had the pleasure of visiting Joann Grohman on her Carthage, Maine, farm, and we got to experience this woman's magic first-hand. We drove for hours through the western Maine woods to get there. Our sojourn was like going back in time to another place earlier in our lives. Here was the complete simplicity and goodness that we had longed for so many years ago. This was a true "cow economy" — a self-sufficient farm based around one dairy cow with chickens and pigs. After tea with warm milk and a tour through gardens of lilacs and comfrey, we left wondering why we had become commercial dairy farmers instead of following our hearts as this woman had done all these years. There was something magical about Joann as she walked barefoot through her realm. Her wealth and experience are chronicled in this volume that has something for anyone with cows.

As we drove away in the early evening, we were filled and inspired with so much awe that we were pulled over by the local police for driving over the speed limit in a local town. We left Joann Grohman's dream world to return to our normal lives. A meaningful life with one family cow might be out of reach for most of us, but this book can help us achieve some sanity in a world that longs for simplicity.

—Jack Lazor

Preface

Every day, for ten months of the year, my cow Fern translates the sun and rain that fall on my small acreage into life-supporting, nourishing milk. With superb efficiency she borrows the energy in cellulose (think grass, upon which humans starve) and reinvests it into a food more perfect than anything in the supermarket. No mangrove swamps or rain forests are destroyed in the manufacture of this product. No water to float a battleship is diverted for her purposes. No grain that might otherwise have been cracked into ethanol or bargained to the starving is apportioned to her use. Fern accomplishes this feat without burning any gasoline. She does this through the magic of wild fermentation in her rumen, the same process employed by cabbage worms and everything else that lives by splitting cellulose. Right now is the moment to abandon the fiction that cows are high on the food chain. The only things that live lower on the food chain than cows and caterpillars are bacteria. Fern is a one-stop food factory, using sun, rain, grass, and rumen fermentation to produce complete-protein milk; the cow does all the work today, and tomorrow she will do it all over again.

Fern's predecessor, old Helen Hefferlump, finally got so arthritic that we knew making it through another Maine winter would be a painful hardship and we needed to end her suffering. Should we wait until she broke her hip on the ice? The options were burying her (impossible in frozen ground) or putting her into the freezer where she would feed many people for a year. Why would Helen prefer to be wasted? Our local butcher said he and his family were raised on aged dairy cows and the meat would amaze us. He was right. Here we can abandon another fiction: the idea that avoiding meat takes strain off the planet. Something eats everything. Life and death merge. Taking animals out of the equation just leaves a vacuum to be filled by insects.

Our prevailing livestock production system is grotesque but at its worst does not remotely approach the waste attributed to it; those mega water requirements and fossil fuel demands attributed to livestock are scary memes, their numbers too vast for mental arithmetic. Efforts to discover any research-based evidence for these beliefs will founder because none exists. And food policy based on fictional numbers is doomed.

Pursuit of an anti-meat agenda will only delay effective investment of our energies. The choice is not between maintaining our omnivore status or "getting over the meat habit and freeing up grain for hungry multitudes," and it never will be.

The cultivation of plants depends upon stoop labor or fossil fuel; apart from animal traction, there is no other way to achieve it. Grazing cows efficiently harvest their own food.

The waste in our existing food production system comes from pulling food out of the loop between soil and eaters, commodifying it, bashing it around to its nutritional detriment, and selling it back to consumers who have already paid for it once with subsidies and will fork over at the cash register and then pay again at the doctor's office. The assumption that glues together the corporate food system is that you and I will not and should not be bothered with home or strictly local food production. I have spent most of my eight decades living the real food truth that, yes, you can produce your own food without any help from agribusiness and with minimal dependence on fossil fuel. Not only can you do this, but it is a source of satisfaction, and often joy. You always know you are doing something worthwhile. The small local farm including animals is the only reliable land-based food production unit. It is a microcosm of the natural world. It creates no environmental debt. It is safe. And it will belong to you.

Acknowledgments

All my family and many friends have provided encouragement with the project of writing this book.

I wish to express my special gratitude to my sons, Dr. Bret R. Luick, Martin J. Grohman and Max J. Luick for scientific and material help throughout.

CHAPTER 1

A Brief History of Dairying

Dairying in its most reductionist form – merely swiping some milk from a cooperative grazing animal – goes so many thousands of years back into pre-history that we can't get a fix on it. It is known that Laplanders herded and milked reindeer eleven thousand years ago. Thirty thousand years ago, people in the High Sinai were confining and breeding antelope with the aid of fences, a human invention arguably as important as the spear. Wherever antelope, reindeer, sheep, camels, goats, or cattle have been brought under human control, they have been milked. The value of milk is obvious and doesn't need to be taught or invented. Among the very earliest human artifacts are vessels containing milky residues.

Horses too have been milked. The hordes of Genghis Khan swept out of Asia eight centuries ago on tough, speedy horses. They triumphed everywhere on account of two important military advantages: they used stirrups, thus freeing both hands to employ weapons. And they had a lightweight, high-protein food source always handy: mare's milk, ingeniously dried by their wives prior to their raids. Each day a horseman put about half a pound of dried milk into a leather pouch, along with a bit of water, and by dinnertime he had a tasty fermented yogurt-like food. No army travels far nor fights well without provisions. Because he didn't have to wait for the quartermaster to catch up, Genghis Khan always maintained the advantage of surprise.

More peaceable folks milked goats and sheep. Sheep and goats have the advantage of being able to thrive on steep, rocky land, and they reproduce rapidly. Gestation takes only five months, they often have twins, and they are old enough to breed by one year of age. But wherever people have the choice, they choose the cow. The cow entered into a relationship with humankind at least ten thousand years ago, and she has been lovingly nurtured and defended throughout Africa, Asia, and Europe ever since. So long ago was she chosen and so much was she valued that her wild ancestors have vanished.

The last known wild cow, called an auroch, died over five hundred years ago in Poland. The cow is the only domestic animal for which there is no wild population pool. Like corn, the staple of the Aztecs that now exists only as farm seed with no wild sisters, the cow now lives in symbiosis with humans.

Archaeologists and anthropologists have shown much greater interest in the role of grains in human history, speaking of what came before as "mere" herding. In fact, discussions of the modern diet seem oblivious to the long pre-history of herding. Arable farming (grain) began about ten thousand years ago. Most writers link arable farming together with animal husbandry, apparently assuming they sprang up together. Many otherwise well-researched sources state that dairy products are a comparatively recent addition to the human menu. To the contrary, grain is the recent inclusion in the human diet, not dairy foods. This false assumption about dairy foods appears to be linked to the widespread belief that milk production is dependent upon grain.

To produce grain in useful quantities requires rich, flat land such as floodplains. It also requires a huge amount of energy, available in antiquity only where complex cultures had developed. Slaves provided this energy. The more slaves you had, the more grain you could grow. And the more grain you could grow, the more slaves you could afford, thus giving rise to a wealthy class able to afford monumental tombs and other durable artifacts of civilization.

Grazing animals have been around for millions of years thriving on grass. They are not dependent on grain. For many thousands of those years they were herded and milked, tasks that require neither slaves nor even permanent dwellings.

Though it is not their natural fodder, cattle will eat grain. In fact, when a grain surplus occurs it can be fed to cattle, which thrive and fatten, as we know very well. This expedient has been practiced from antiquity and continues to this day in the American Corn Belt. If the price of corn falls, farmers may decide to "put it through cattle" instead of selling at a loss. In impoverished parts of the globe the animals then become walking food storage units, later to be eaten or, as in the United States, to be sold at auction.

Herding animals requires only the availability of shepherds and can be done on any kind of land from rocky mountainsides to kelp-strewn beaches. Wherever herbivores have been herded, their milk as well as their meat became an important part of the local diet. Herbivores convert grass, bushes, and weeds into high-grade, readily available food. They do this with enormous efficiency whether in captivity or not. But when herded, they free up a great deal of human time for other pursuits. Humans with extra time and energy tend to engage in commerce, the arts, invention, and war, not necessarily in that order. Dairying has played its part in these pursuits.

Dairying also led to the use of the fence, which in antiquity served less to keep animals from running away than to protect them from predators at night. Ancient Sumerian writings reveal that fencing also provided a means for keeping the best milk-producing animals close at hand. But fencing forfeits the transcendent advantage of grazing animals: that they find their own food. It is only feasible where there are servants to fetch and carry feed to the animals. Like grain production, the fence is evidence of a complex hierarchical culture, and both are natural by-products of the civilizing influence of dairying.

The fence served another function basic to animal husbandry: it permitted selective breeding of cattle, sheep, and goats. By confining smaller and more docile males and permitting only these to breed, at least ten thousand years ago people were manipulating animal genetics. Domestic breeds began to have smaller horns and to be of a more manageable size. This was particularly important in the case of cattle, which like all dairy animals were often handled by women and children – the wild cattle were huge and quite dangerous. Although in actual numbers worldwide there have always been more sheep and goats being milked than cows, with the selective breeding practiced by early cultures, the cow very quickly became the most prized dairy animal.

The Cow, Premier Dairy Choice

The cow is the premier dairy animal because of her cooperative temperament, the comparative ease with which she can be milked, the volume she is able to produce, and the versatility of her milk. The cream is easily skimmed and made into much-prized butter in cold climates and ghee in hot climates. (Ghee is butter that has been heated and strained.)

The cow is a primary producer of wealth. She can support a family. She not only turns grass into milk in quantities sufficient to feed a family but also provides extra to sell *and* contributes a yearly calf to rear or fatten. The by-products from making cheese (whey) and butter (buttermilk) will support a pig or two. Her manure improves her pasture and when dug into the garden results in plant growth unsurpassed by other growth mediums. The family that takes good care of its cow is well off.

Cattle are the original stock in "stock market." Ownership of cattle has always been a marker of wealth. This is not just because the cow is a primary producer of wealth, adding enormous value to grass. In a "which came first, the chicken or the egg" sort of way, it's also because only families possessed of a hardworking, cooperative spirit are able to keep a cow, let alone build a herd. Cows require humans for their survival.

Other domestic animals can revert to a wild (feral) state with predictable success. Put hogs in the woods and they hardly look back. They won't get fat, but they will immediately form a breeding population. So will horses on the plains. Many breeds of sheep can establish themselves in hill country. Goats are well known for this aptitude (so long as they are not too far from the sea; they have a high iodine requirement).

So Huckleberry Finn's Pap might have had a pig or goat he could turn loose and still call his own, but a cow requires consistent responsible care. If she doesn't get it she won't give milk and she won't start a new calf and she won't live through much cold or drought. Farmers in the north put up hay for the winter. African herdsmen walk their cattle to water and defend them against lions. Even the great beef herds of South America and the American West have been dependent on humans to arrange things for their benefit; the wolves and mountain lions didn't disappear by accident. But this story is about the dairy cow.

The dairy cow doesn't ask for much, but she asks every day. People who are creating wealth with a cow either are hardworking and reliable or get that way in a hurry. This is the way it has been for a very long time. The fine farms of Europe, England, New England, and much of the rest of the United States were all established thanks to the wealth derived from cows. Wherever there is, or used to be, a big barn, it was likely built to store winter hay for the cows that once dotted the pastures. The need to milk the cow twice a day determined the location of many a church; people had to be able to walk there and back without disruption to the schedules of cows. Formerly, every district in Europe, England, and the eastern United States had a corn mill situated so that a farmer driving a horse and wagon could deliver his load and get home in time for milking. It is certainly no coincidence that such a large number of our finest statesmen were born on farms. Important virtues are nurtured on the farm, including a graphic understanding of the relationship between working and eating. Over my farming life I have bred and raised all of the traditional farm animals, and I love them all. But through association with the dairy cow I have come to understand

and accept the words of that great nineteenth-century agricultural essayist William Cobbett: "When you have the cow, you have it all."

If Cows Are So Great, Why Doesn't Everybody Have One?

Not so very long ago, a great many people did indeed keep a cow, and she was often an adored member of the family. Well-to-do families even in cities kept a cow well into the early part of the twentieth century. During the Victorian era, country homes of the wealthy included charming accommodations for their cows. Some of these were quite fanciful and included beautifully tiled dairy rooms for making butter and cheese. All this attests to the high regard in which the dairy cow and dairy products were held.

Peasant homes were built to take advantage of the considerable heat given off by a cow. In Scotland often the cottage was built to surround a stall in which the cow spent the winter; picture an arrangement like a playpen in the middle of a low-ceilinged room. In other locales, including Spain, the family lived in rooms above the cow byre, using cows like a furnace in the basement.

Some of the forces that led to the demise of cow keeping were the same ones that have stressed the American family. The automobile was important; it dispersed families and directed interest away from home-based activities. A desire for consumer goods that can't be satisfied without focusing the whole energy of the family on acquisition was certainly a factor. That consumerist mind-set fostered a yearning for enhanced social status. We belong to a hierarchical species; our need for status is hardwired, but its expression is culturally determined. There have been eras where the cow accords status, and there still are places where that principle holds true. But nowadays status is more likely to derive from real estate in a good location. On a rural property, the high-status animal is the horse. But all these factors are as chaff compared to the power of the twentieth-century revolution in food production, processing, and distribution.

The food revolution is lauded in school texts, political speeches, virtually everywhere as an exemplary modern triumph that has showered us with endless choice and abundance. Nevertheless, warning bleats from people like me, pointing out that our food system is wasteful and nutritionally compromised, are mounting in volume. The most astonishing feature of this food system is usually overlooked: food has been commodified. We take that commodification for granted; it's embedded into our cultural psyche. But for all of human history until very recently, and in a few places still, food is something you find, you grow, you fish from the sea, or you obtain locally from the actual producer. The purpose of this food is straightforward and obvious: it is to feed people. If sold, it changes hands only once. It goes directly to the people who will eat it. Designer food intended only as a source of profit has arrived late in humankind's history.

The foods in our shining supermarkets were produced as a financial investment. They are not so much food as consumer goods. As such, the primary constituents of the majority of the finished goods – the wheat, corn, edible oils, soy, and sugarcane or sugar beets – are grown as a monoculture on millions of flat acres, the constituents broken down and reassembled into something that keeps nicely and resembles food. At each of the many steps in production, value is added. Yet the only entities to which this value yields profit are the corporations that manufacture the food products.

The Commodification of Milk

Because of its extremely perishable nature, milk initially presented a challenge to commodification. In the late nineteenth century, as American cities began to rapidly expand, the demand for milk was met in several ways. One enterprising solution was to position a great barn full of cows right downtown next to the inevitable brewery. The cows were fed the spent malt. In theory this production practice could have proven satisfactory; in practice it was disgusting. The cows were kept in filth and were milked by hand by anybody off the street. On top of that, the milk was routinely watered. Rural dairies had a better reputation and made a valiant effort to get milk delivered fresh and cold by train.

Milk trains moved through the countryside before dawn picking up the familiar milk cans that waited on platforms. The milk did not travel great distances, and it was bottled and delivered fresh to doorsteps every morning. Cans were kept cold by blocks of ice cut from northern lakes in winter, where ice cutting was an important industry. The big blocks were packed in sawdust, available in quantity from sawmills, and kept right through the summer.

Dairymen well understood that milk quality depends on healthy cows, clean milking practices, rapid chilling, and expeditious delivery. Milk itself tells the tale at the table just as unmistakably as does fish. There are two ways to achieve a safe, edible product. The first is by conscientious handling. The second is by sterilizing and preserving the milk (or fish or any other food), after which it matters a great deal less how it is stored or for how long. Small dairies able to exert quality control every step of the way, often even bottling and delivering their own milk and cherishing the one-on-one relationship with their customers, supported the first method. Larger, well-funded consortiums seeking control of dairying favored the second. Their approach was to pool larger quantities of milk, drawing it from greater distances and overcoming problems of quality by heat treatment. They called the heat treatment *pasteurization*, tapping into name recognition of the great French scientist Louis Pasteur.

The outcome of this disagreement – conscientious handling versus sterilization – was by no means a foregone conclusion. Heating changes the appearance, flavor, and nutritive and culinary properties of milk, and none for the better. As for its keeping qualities, everybody and his grandmother knew milk goes sour after a few days. It wasn't expected to keep; after all, that's why we make cheese. Everybody preferred fresh milk and understood perfectly well that pasteurization served as a substitute for quality. Dairymen who wanted to continue selling fresh milk geared up for more efficient delivery using ice and seemed about to make their case for quality control at the source. Quite apart from concern for their customers' preferences, this enabled them to maintain financial control of their own product.

Then came the winter of 1886, when the lakes didn't freeze. Lacking ice, the case for fresh milk was lost by default. Dairy farmers were forced to sell their milk to middlemen, as they do to this day. They have never been able to regain control over their own product.

Consumers had their minds changed about pasteurization by a fear campaign promoting the disease hazards said to be unavoidable from unpasteurized milk. Indeed, such claims are likely to be true for the modern dairy system, which pools milk from thousands of cows at a time. Although, then as now, it is perfectly possible for herds to be clean and disease free, what is *not* possible is for that milk to stay fresh when it is transported great distances and left on the shelf for weeks at a time. Pasteurization was instituted for the benefit of distributors; it keeps milk from going sour and becoming unsalable. But a nervous public accepted an array of new public health statutes fostering the concept of pasteurization. Indeed, America was in the mood to sterilize everything possible. It was the heyday of the hospital-white kitchen and bathroom. Dairymen were required to paint everything white too, as part of the mystical association of whiteness with health and cleanliness. To this day, dairy farmers must conform to public health regulations far more strict than those imposed on any other industry, including the very processing plants where milk is conveyed to be pasteurized.

What Happened to the Family Cow?

What was happening to the family cow while commercial dairying was being conformed to the twentieth-century model of food as commodity? Along with small farms of every sort, she was being priced out of existence. If you talk to old-timers you'll hear them say, "It doesn't pay to keep a cow." American food is cheap, or at least appears to be. Starting early in the twentieth century, an elaborate system of subsidies has kept food prices artificially low, part of a cheap food policy that favors city dwellers. This policy has been continued by every administration. Most people are familiar with farm subsidies, payments to farmers that assist them in producing at predictable levels. Among the less frequently recognized effects of subsidies is the fact that because they cover part of farmers' costs, they enable farmers to accept a lower price for their crops or milk so that we can pay less for food. Pointed out even less frequently is the program of government assistance for processors. Everything from special university research projects to tax-deferred production plants may be paid for wholly or in part by tax dollars. Highway costs are shared by all taxpayers but benefit truckers – and the food industry – disproportionately. This is sometimes referred to as corporate welfare. These are some of the hidden costs of cheap food.

With food costs comparatively low, even the formidable efficiency of the cow is hard put to offer an obvious fiscal advantage. Milk prices remain low because dairy farmers are paid at a rate that barely covers costs and they cannot market their milk freely. They must sell to consortiums under fixed contracts that are regulated by the government. And processors have certainly made milk conveniently available in markets. If a plot had been hatched to eliminate small farmers, place milk production and distribution in the hands of the few, and permit almost everybody to forget what milk was meant to taste like, a better plan could not have been devised. Consider also that in terms of buying power, American wages were high during the first sixty years of the twentieth century, when these government subsidy programs were put into place, and even today our dollars still buy more food than in any other Westernized country. So keeping a cow might indeed cost more than buying milk at the store, especially if the cost accounting is narrowly focused. (The up-front costs and prospects for profit of the family cow are analyzed in chapter 19, "Your Cow Economy.") Most people considering a family cow are no longer motivated by the old-timer view that the object of a cow is to "pay," reasonable as this may be. Instead, they're often looking for quality dairy products and a more centered way of life.

Treatment of Commercial Milk

Pasteurization has its critics and I am among them, but without it, there is no doubt that milk distribution as we know it would be impossible. Pasteurization destroys all bacteria, including benign (or even beneficial) strains, and it destroys enzymes, besides physically altering milk protein. In addition, since the 1950s virtually all milk has been not just pasteurized but homogenized. Since most people have not had the opportunity to become familiar with milk in its natural condition, these two terms, *pasteurized* and *homogenized*, which appear on every container of milk in grocery stores, have become confusing. *Homogenized* means the milk is subjected to pressure and agitation that knocks apart the butterfat globule and stops it from doing what cream would naturally do: rise to the surface.

Homogenization, too, was presented to the consumer as a great advance but first and foremost served the distributors. Before the days of pasteurization, cream was prized, but after heat treatment cream becomes lumpy and unimpressive. Homogenization offered the advantage of distributing the cream evenly throughout the milk "so everyone gets their share." The advantage for distributors was less charming. Because pasteurization made it possible to sell milk two or even three weeks after it left the farm, there emerged a problem with a sort of sludge settling to the bottom of the bottle. (Don't let me put you off store-bought milk altogether; worse things are found in ketchup and peanut butter.) This sludge consisted of dead bacteria, somatic cells, and the macrophages that consumed them, and the longer the bottled milk sat, the more evident the sludge. With homogenization the sludge became invisible, along with the cream.

This might seem reason enough to get a cow, but now we have bovine growth hormone, or BGH, to consider. Consumers have expressed virtually unanimous objection to the fact that milk may now legally contain BGH that has passed into the milk as a result of the cow's daily injection. The history of milk distribution does not offer much reassurance that our concerns will end the practice.

BGH (rBST, rBGH)

In the dairy industry these days, there's a lot of talk about rBGH (recombinant bovine growth hormone) or rBST (recombinant bovine somatotrophin); these are the names given to the genetically engineered version of bovine growth hormone. It is now being used in every state. The U.S. Food and Drug Administration (FDA) has declared milk from cows injected with rBGH to be safe and has advocated making it a federal offense for processors to so much as state on milk labels whether or not the milk is from rBGH-treated cows. Dairy farmers hoping to serve a niche market of discriminating consumers desiring rBGH-free milk offered legal challenges to this federal directive and were defeated. The courts found that any claim that milk is free of rBGH could be interpreted as implying that other milk produced through the use of rBGH is less safe, thus contravening the government's position. The FDA has since relented sufficiently to permit dairies to label milk as BGH free, but they must also add a disclaimer stating the FDA position.

The stated objective of BGH treatment is to cause cows to produce more milk. That's what it's for. The value to the dairy industry resides in the ability of rBGH to induce cows to produce more milk than they had meant to . . . whether they want to or not, so to speak. This point may seem painfully obvious to those who assume that so-called factory farming (now usually referred to as confined animal feeding operations or CAFOs) already includes dairying. It does not. Factory farming is the practice whereby hogs and chickens live out their lives in such close confinement that sows cannot even turn around to see their piglets, which must suckle under a steel barrier, and a hen can't even stand up and flap her wings after her egg rolls away down a conveyor belt. Factory farming is more profitable than traditional farming because less human labor is required and unskilled labor will suffice. Factory farming has become an investment opportunity, and traditional producers of pork, chicken, and eggs have everywhere been driven out of business by these aggressive giants.

Modern cows produce a great deal of milk, some exceeding ten gallons a day, but they do it as a result of selective breeding over many generations, much the way racehorses have been bred for speed. And even then, a cow's milk production, like the racing success of the horse, cannot be expressed without scrupulous care and feeding right from birth. Whether the dairy cow lives her life lined up with a thousand others or the barn holds only thirty – or just one – she has to be kept happy. To stay in the mood to produce lots of milk she has to have appetizing food, fresh air, and a comfortable area to stand and lie, and she must be politely treated. Otherwise she won't produce to her genetic potential. Otherwise she won't let down her milk.

With these prerequisits for humane treatment in mind, dairying has so far avoided the worst characteristics of factory farming. Will rBGH force cows to produce milk around the clock while perpetually connected to a milking machine? Not exactly. In fact, dairymen using rBGH report the need for added personal attention to the cows because they become fragile. The rBGH does, however, compel the cows to divert more of their energy to milk production. It may be compared with drugging a racehorse to eke that extra burst from an animal already bred to run with speed and courage. BGH does move cows in that direction. As has already occurred with pork, beef, chicken, and eggs, huge centralized production facilities – let's not call them farms – lead inevitably to bland, standardized food with no added assurance of safety. This is food obtained from animals living on feed compounded from by-products often so disgusting that euphemisms are invented by the industry to conceal their identity. Have you ever looked at a dog or cat food label and wondered what "poultry digest" might be? Chicken manure mixed with litter is also permitted in dairy feed.

A cow's production peaks at about six weeks after calving and then begins a gradual decline until it is time to dry her off in preparation for her next calf. The use of rBGH prolongs the period of peak production, when nature would have her diverting feed to maintenance and to her fetus. Additionally, in order to speed cows through the milking line, commercial dairies often use oxytocin injections. Normally, cows require a few minutes of preparation before letting down their milk, and if they are stressed they will let down incompletely or not at all. Oxytocin causes immediate letdown.

That is rBGH from the perspective of the farmer and the cow.

Health Problems Associated with BGH

BGH was first marketed by the Monsanto corporation. Recent information from Canada, where Monsanto attempted to gain approval for the use of rBGH, must cause us to reconsider the claim of total safety that the FDA has made for BGH. This claim was made based on the representation that rBGH is broken down in the gut like any protein and does not enter the bloodstream, and thus cannot cause harm to the consumer. Canadian government scientists who reviewed Monsanto's rat and cow experimental data concluded that the hormone does indeed enter the bloodstream, as evidenced by an elevated immune response. In rodent studies, there were reportedly cysts on the thyroids of male rats and damage to the prostate.

The Canadian scientists also reported that IGF-1 (insulin-like growth factor 1) is elevated in the milk of cows injected with rBGH. The growth hormone IGF-1 is identical in cows and in humans.

The Cow as Security

Does the future seem uncertain? One of the best ways to take charge of your own future and that of your loved ones is to grow your own food. This is a life-affirming choice and one that may well offer better odds than going about armed to the teeth. Inasmuch as this is hardly a new idea, many schemes for self-sustaining food systems have been devised. One method involves growing algae in vats on the roof. Another promotes earthworms and other insects as ideal basic food. There are systems for backyard fish ponds capable of growing many pounds of a fish called tilapia by adding manure and other waste to their water. Some people advocate planting family-size patches of soybeans along with other vegetables. All of these approaches offer food security of a sort, along with major problems. Algae taste awful, insects don't appeal to the Western palate, tilapia are nourishing but boring. An all-vegetable diet is seriously boring and is extremely labor intensive. Furthermore, the oft-heard claim that the all-vegetable (all plant, no animal) diet can provide complete nutrition is not backed by research evidence. I recommend the cow.

Maybe We Should Just Start with a Few Chickens

If the biggest animal you've ever known personally was a golden retriever, the cow may seem like a giant step into the unknown. Why not start with a pig, then, or maybe a few chickens, or a goat? Those are all fine animals, without doubt. Pigs are easily kept in a small space and provide a carcass with a high proportion of meat to bone. There is very little waste and everything you do with the meat is delicious. One consideration is that the food preferences of the pig are so similar to our own that in a very real sense it competes directly with us for food. Some historians believe this to be the reason wise leaders in some cultures, including the Muslim and Jewish ones, have proscribed pork. Nevertheless, in circumstances where there is often food that would otherwise be wasted, sometimes called default feed, the pig can be very rewarding.

Chickens are wonderful, and I would not be without them. They are extremely easy to keep, and I love having my own eggs, even though I rarely get up the gumption to kill one of the birds. Consequently my return on investment is diminished by having to maintain an aging population. But perhaps people are saying this about me.

Goats are appealing and with good management highly productive. In terms of production it takes about five goats to equal one cow. They must be penned behind excellent fence, or tethered. Unless penned in a huge area, because of their agility and eclectic tastes, they will soon have the ground reduced to a desert; then their food will have to be brought to them. Goats are very active. If tethered they must be well supervised to avoid entanglement and moved frequently to fresh browse; otherwise production will plummet. In rough, brushy terrain where there is no grazing, goats are ideal. But a goat is not easier to take care of than a cow, and cow's milk is considerably more versatile than goat's milk.

Amazing Cow Magic That Most People Don't Know About

"A young fellow wantin' a start in life just needs three things: a piece of land, a cow, and a wife. And he don't strictly need that last." That's an old saying that once annoyed me. Now that I'm an old lady with a cow but no spouse, I am prepared to concede at least the validity of the underlying premise.

An overarching truth about the cow is that she drives the domestic or small farm economy. She lives on a constantly renewing resource: grass. Her rumen microorganisms assemble complete protein from which she makes milk. With this milk she is able to not just provide for her calf but also meet the protein requirements of your pig *and* your chickens (neither of which can live on grass) and still provide milk for the house. The reverse is never true. No pigs or chickens *or any other nongrazing animals* can live on grass alone . And the cow does it on free resources: water, soil, and sunshine. Through her sovereign ability to convert grass, which otherwise has no value, to milk and meat, which do have value, the cow produces wealth. She thus vaults the domestic or farm economy into a self-sustaining mode. Even with the most exacting sweated labor in the orchard or garden, you can't grow plants that will support reproduction in pigs or chickens or any other nongrazing species, including humans; you can fatten them, but growth will be retarded and fertility negligible. This key fact about cows should never be forgotten.

The fact that fruits, nuts, grains, beans, vegetables, grasses, or any combination thereof will not support reproduction in nonherbivorous species has been forgotten by urban dwellers. Seemingly well-informed people appear confident that somewhere, somehow, growth, reproduction, and vigor can be sustained directly via the soil without recourse to animal intermediaries. Such beliefs would not survive a year on a farm practicing sustainable agriculture. Those few urbanites who have committed themselves to the attempt to feed themselves on a home-grown diet devoid of animal products find it exhausting. But add a cow and you no longer need to push nature uphill. The cow does the work of converting plant products to food of unassailably high quality. Fiber, upon which single-stomach (monogastric)

Circle of sustainability

species such as ourselves would quickly starve, is her preferred food. She thus serves as a bridge between the plant world and ourselves and all other monogastric species.

Centering

At various times I've kept a cow in a suburban garage and grazed her on the lawn. It definitely can be done. But let's say for now that you have a cow on five acres. Let's also assume you really appreciate good food – its flavor, its variety, the way your friends rave about the quality. You also like knowing where your food came from and what *isn't* in it and on it. Perhaps you've also read some old Adelle Davis books and have recognized that good health isn't something left over after you eliminate all the risks. It's something you build with real food, such as milk and meat and all the fresh fruits and vegetables that thrive on the composted cow manure you have out back. Probably you aren't the son or daughter, or even the grandchild, of one of those old-timers who say cows don't pay. Maybe, like me, you like to work at home, trading time back and forth between the computer and outdoor jobs.

This way of life is centered and rewarding. If there are children in the mix, they will be robustly healthy and will develop beautiful bone structure. And since it will mean one or both parents can be at home, they will also be happy. Later I'll talk about how and where the money sorts out in this way of life. But of this you may be certain: it's worth it. If you have plenty of money, it's a worthy commitment, and you will be the envy of your friends. If you have very little money, it offers magic-carpet access to a lifestyle currently unattainable to most working couples, who often feel forced to choose between having children and paying the mortgage. It makes possible a very good life on one paycheck.

No Land?

"I don't have a farm. I just want my family to have healthy milk. So, can I keep a cow in my backyard?" Yes, you can keep a cow in the space you might need for a swimming pool. You can keep a cow on a large lawn and dispense with the lawn mower. This is literally true, and I have done it, but to manage flies and maintain the tidy appearance of your property, somebody will have to go around every day with a shovel and put the manure on the compost heap. So long as there is grass, cows will ignore most ornamental plants;

however, they will lie down and mash them. A cow must have longer grass than do sheep or goats, which have a split upper lip that enables them to graze turf very closely. Cattle, sheep, and goats all lack upper incisors. Horses have upper incisors and will completely devour pasture right down to the dirt. They can also bite, something ruminants cannot do.

Now would be a very good time to get a cow even if you have only a backyard. The quality of all purchased food is declining in nutrient value while merely appearing to increase in variety. There is little reason to believe it is getting safer. Fresh milk adds a life-supporting dimension to the diet unmatched by any other possible food. Even without other dietary improvements, drinking fresh milk will bring about a striking improvement in family health. I have seen this many times.

Let's Get a Cow

Some small dairies catering to niche markets will survive to serve the discriminating customer. There is a rapidly growing market for organically produced milk and cheese, for example, but the higher prices discourage many customers. And in most states dairy farmers who sell milk directly to consumers face crushing fines. They must sell to processors or throw their milk away.

For an increasing number of people now, the answer is, "Let's get a cow."

Cow Futures

Is there a future for the cow, or is she an anachronism, a leftover from a historical era best forgotten, albeit an era that has lasted perhaps thirty thousand years? Dairying as now practiced will continue its current trend toward corporate farms milking over a thousand cows. Smaller family farms will continue to disappear at their current alarming rate of over one thousand per month nationwide, with few young couples entering dairying. Dairy farmers have shown little aptitude for defending the reputation of milk against an onslaught of disinformation about its health properties, despite fully adequate supportive research. Health professionals obsess about fat and allergies to a degree that I find irresponsible, since they ignore the developmental outcomes of eliminating dairy products.

Dairy farmers own their cows but have no influence over the fate of the milk. The powerful food processors and marketing groups that own the dairy products lack incentive to set the record straight because real dairy foods are in direct competition with more profitable imitations that they also market. Among these are margarine, coffee whiteners, whipped toppings, fake cheese, vegetable oils, and milk by-products such as casein and whey. These and all of the many soy foods now prevalent in the marketplace not only compete directly with dairy products but are more profitable. Relentless promotion of low-fat milk is backed by dairy manufacturing interests that market cream more profitably in ice cream. The quality of commercial dairy products cannot be expected to improve under these conditions. Therefore new customers are few. The sum of these forces will ensure that the use of commercial milk will continue its decline.

Making Our Own History

None of this need affect us. If you now have a cow or are considering one, you probably already appreciate her advantages. Further reinforcement for the pro-cow position will be found throughout this book, but if you keep a cow you will sometimes have more milk than any family could drink. This brings me to a philosophical point. Americans are very conflicted about excess. On the one hand, we are a wasteful people, taking long showers, driving two blocks when we could walk, and allowing more than one-third of all purchased food to end up in the garbage. On the other hand, we are abstemious, and many of us make a point of serving reduced-fat foods and avoiding many traditional favorites. We feel guilty about food and apologetic . . . "Oh, I shouldn't . . . well, perhaps just a tiny piece." Food authorities constantly reinforce food guilt. As surely as reputable research dispels concerns about fat as a factor for one disease, the last paragraph of the report will quote another authority: "But this should not be interpreted to mean you can start eating [whatever] again, because there are still plenty of other good reasons not to eat what you would like."

Real dairy products are now included in what has become a blanket condemnation of fats. In 1989 Thomas Moore published *Heart Failure*, in which he detailed three decades of political (not scientific) infighting among scientists that resulted in cholesterol and saturated animal fat being fingered as the major culprits in circulatory disease. By the time Moore's book was published it had already become clear to researchers that the case against animal fat was shaky. At about this time evidence against fats from vegetable sources reached critical mass; most of them contain either carcinogenic factors (free radicals) or trans fatty acids (dangerously altered molecules introduced by processing). Unlike animal fat, many vegetable oils contain naturally occurring toxins such as gossypol (in cottonseed and canola) and erucic acid (in canola). They may also contain industrial solvents from the extraction process. We did not hear authorities apologizing for the error, saying, "Oops, we got all that fat advice backward." Instead, there was an immediate campaign to persuade the public to simply avoid *all* fats.

A False Economy

As a nation, we are not getting thinner. We are just feeling guiltier. So if eating is something to feel guilty about, then it must be virtuous to spend less on food. The family table is thus being made to bear the burden of the shrinking dollar. Other costs of living are often nearly intractable and little virtue attaches to their reduction. Housing, transportation, insurance, and tuitions require huge nonnegotiable sums. Food expenditures are expected to shrink to accommodate these fixed costs, and nutritionists who ought to know better offer carbohydrate substitutes for nutrient-dense meats and dairy products. Books and newsletters encourage strategies for thrift by authors who parrot popular nutrition ideas but know little about food except what it costs and perhaps how much fat is in it. Health outcomes are ignored in favor of immediate savings. I hate to see waste, and I don't mind clothes from the thrift shop. But stocking the pantry with soy milk and macaroni dinners is a tragically false economy. Imagined reductions in fat are illusory; such food policy results merely in substituting cheap and unhealthful manufactured fats for better natural fats. It leads also to carbohydrates as a protein source. From these inferior protein sources the extra calories cannot be trimmed.

A Psychology of Abundance Is Healthier and More Fun

No need to be stingy. With my cow, I am able to serve exceptionally fine food and I am not stingy with the butter and cream. My cow supports my chickens, so I always have eggs. She also gives us a calf, so there is no meat shortage. Her extra skim milk supports a pig, so there is no bacon and sausage shortage. And she provides fertilizer for the garden, so there is no vegetable shortage. The cow is a generous animal. She improves life for everybody.

CHAPTER 2

What Makes Cows So Important?

> *But raw milk from a Jersey cow is a totally different substance from what I'd thought of as milk. If you do not own a cow or know someone who owns a cow, I must caution you never to try raw milk straight from the teat of a Jersey cow, because it would be cruel to taste it once and not have access to it again. Only a few people in America remember this type of milk now, elderly people, mostly, who grew up with a cow.*

So wrote Kristin Kimball in *The Dirty Life*, an account of her evolution from city girl to farmer. (In fairness, I will state here and now that other breeds are also capable of providing this transformative experience.)

Nature's Most Perfect Food

It still is. It has to be. Like all high-quality perishable foods, milk is best when fresh. Everyone is aware of the difference between fresh fish and fish kept for several days, even on ice. We all know the difference between fresh bread and the stale loaf. Anyone who has the opportunity to compare new-laid eggs with those kept in cold storage invariably notices the favorable difference. Few are now aware that the difference between fresh and store-bought milk is of the same order. Home-produced milk varies in flavor due to seasonal and other factors, but you can count on a light, delicious, and fresh taste unknown in commercial milk. Store-bought milk has a cooked, lifeless taste with chemical undertones. The processing industry knows this but so far has dismissed criticism by saying, "People have gotten used to the taste of plastic." I am not making this up.

If enough of us become reacquainted with the superior flavor and other important qualities of fresh milk, cream, and butter, processors will respond by improving quality and offering more choice in commercial dairy products, just as has occurred with bread. People in the last few decades, in open rebellion against the dismal quality of commercial bread, began baking at home. Commercial bakers quickly responded by offering better bread, and soon excellent niche bakeries sprang up.

There is now organically produced milk available commercially, and the brands I have tasted were markedly better in flavor than standard choices. At present the demand for organic milk is driven by a desire for milk from cows that have not received antibiotics, hormones, or feed supplemented with recycled waste. Surveys tell us that society's current food goals are health and safety, the environment, animal care, and sustainability. Flavor has not so far entered the discussion. Organic dairies must perforce use harsh sanitizers, and the milk is sold in the same plastic-lined cartons as other milk. It is typically pasteurized and nearly always homogenized. The enzymes are destroyed and the fat globules cracked apart just as completely as in common commercial milk. I very much applaud the dairy farmers who produce milk organically and the families who buy it. It is a giant step forward in a renewed commitment to family health and family farms. Untreated (raw) milk in opaque glass bottles would be a logical next step to preserve both flavor and nutritive value. But unlike the consumer decision to make bread at home, there is a vast bureaucracy with beliefs chipped in stone defending the pasteurization (heat treatment) of milk. (There are no statutes requiring homogenization, but unhomogenized milk, heated or not, travels poorly. The cream gets lumpy, or even turns to butter.)

Pasteurization

Pasteurized milk is heat-treated milk. Pasteurization is achieved by one of these three methods: heating milk to a rolling boil for one second, to 170°F (77°C) for 15 seconds, or to 150°F (66°C) for 30 minutes. The last method best preserves the flavor of the milk. Pasteurization inactivates all enzymes and destroys bacteria. Much milk is now ultrapasteurized, by heating it to 280°F for two seconds, which gives it a shelf life of at least six months.

If you serve your fresh milk to people who have not given the matter much study, you will soon encounter the prejudice against unpasteurized milk. Raw milk has been virtually unavailable in this country for three generations. Hesitancy to use it now amounts to a classic food taboo in that avoidance is both irrational and emotional.

No American child gets through school without encountering textbook material in praise of the pasteurization that "makes milk safe to drink." No science course omits praise of Louis Pasteur, the father of bacteriology, the man who made possible modern medical practice. Those supporting pasteurization are able to take comfort in the assured approval of an important orthodoxy. Conversely, to question the necessity for this treatment is to be placed in a sometimes awkward position. One needs to understand the behavior of milk and of the bacteria that may (or may not) inhabit it in order to explain to critics the circumstances under which pasteurization is valuable and when it is destructive and unnecessary.

For the purpose of current commercial distribution of milk, pasteurization is an undoubted necessity. Not only does the dairy farmer pool milk from many cows under circumstances that make it impossible to avoid occasional contamination with fecal material, but this milk must be kept in a holding tank with milk from previous milkings, and it is warmed twice a day by the addition of new milk.

In some areas milk is collected only once a week by the milk tanker, although the more common practice is every other day. The tank driver is required to reject milk that is above a specified temperature (usually 45°F). In practice this is not always done, since some drivers do not have the heart to tell the farmer he or she must discard his or her milk when refrigeration fails. The driver knows the dairyman is already farming on the edge.

In terms of flavor, pooled milk even before it leaves the farm does not compare favorably with what you milk into your pail by hand. When I had a commercial dairy herd, I attempted to get around this problem for my own family by having the milk from a selected cow go straight from the milking machine into the household milk can; even this was not quite up to the quality of hand-milking into a bucket. The entire milking system must be washed with powerful disinfectants daily, and a trace of this finds its way into the milk. (When milking with your own one-cow machine the flavor is better protected.)

Most dairy farmers are meticulous and conscientious, and organic farmers especially so. But a few are not, and we all know examples of the difficulty of correcting the behavior of an offender in any field. It can be slow going. Some farmers, being warned that the bacterial count in their tank is creeping up, will pour chlorine bleach directly into the milk. There are some, despite warnings, who continue practices that allow cow dung to get into the milking machine. In some states there are dairies that milk several thousand cows and employ unskilled laborers with little personal stake in milk quality. Ultimately everybody's milk joins company in the tanker. Transporting milk for long distances alters the flavor by speeding rancidity, or the oxidization of fat, which releases a cascade of free radicals. Agitated milk also goes sour more quickly due to the breakdown of lactose. Once the milk reaches the processing plant, it is further blended with milk from other tankers, providing a new field day for whatever bacteria are present. These are intractable problems, impossible to circumvent given today's consolidation of dairying.

The bureaucratic empire that defends milk safety is understandably sensitive to the possibility of the outbreak of any milk-borne disease. Some once-common diseases are now rare. Brucellosis (also known as undulant fever or Bang's disease) is one. Testing programs continue, but most states have been declared brucellosis free. If you were purchasing a cow in a state that is not brucellosis free, certainly no seller could object to your obtaining a test. Few states continue to test for bovine tuberculosis, as it is now nearly unknown in this country. Those states that border Mexico are at some risk, because Mexico continues to have infected cattle. A herd right on the border could conceivably become infected. If there is doubt in your area, testing can be ordered, though in most places your vet will tell you not to bother.

Some less dangerous organisms will always remain a concern because they are dirt- or saliva-borne. The bacteria *Campylobacter jejuni* has been implicated in a number of outbreaks of food poisoning that were traced back to raw milk; in more than one case schoolchildren were given samples of milk from the bulk tank during a field trip to a dairy. Many became ill with gastroenteritis, though none of the farm families were affected. As the *Journal of Food Protection* stated in 1983: "A wide variety of warm-blooded vertebrates have been infected by C. jejuni. This organism has been isolated from the intestinal tract and feces of humans, cattle, goats, sheep, pigs, chickens, and some wild birds. It has also been isolated from poultry meat, salt water, and fresh water. Rapid cooling of milk should prevent growth of C. jejuni and either cooking or pasteurization can be expected to kill it" (vol. 46, p. 7). Note that the organism resides in the gut. If found in milk, it is due to contamination from manure. Its presence in poultry meat results from messy high-speed slaughtering practices.

Salmonella and many species of staphylococcus and streptococcus can contaminate milk, whether from mastitis in the cow, dirt falling into the milk, or careless coughing by the milker. They can reach dangerous levels if the milk is not promptly chilled. These contaminants can just as easily enter milk and other dairy products *after* they are pasteurized. In fact, the majority of disease outbreaks reported in milk and milk products originate from ice cream and fresh cheese, not raw milk. In most years reported outbreaks due to dairy products are in the single numbers, while cases attributed to chicken number in the thousands. Salmonella poisoning can have serious consequences, including death in the very young, aged, or infirm. It is usually salmonella that sends people to the hospital after they eat inadequately cooked turkey stuffing or handle commercially slaughtered raw chicken.

E. coli O157:H7, an aberrant strain of a normal gut bacterium, has been identified in some commercial dairy herds, although it is more characteristic of feedlot beef cattle. If found in milk it is due to fecal contamination. Its presence in the bovine gut is closely associated with high rates of grain feeding.

A number of other communicable diseases affect cattle and can cause severe loss of production (see chapter 16, "Diseases and Disorders") but do not affect humans. I have not encountered them in my own cow and rarely had any reported by my readers. Bovine spongiform encephalitis (BSE) or mad cow disease is discussed in chapter 16. Assurance of freedom from this and other diseases may be one of the reasons you have chosen to keep a cow.

All this makes it clear why commercial milk needs to be pasteurized. Otherwise it could be unsafe and would certainly be unusable.

When it comes to raw milk that is produced under hygienic and humane conditions, chilled promptly, and consumed shortly thereafter, pasteurization is unnecessary and, in fact, to the detriment of the consumer, who loses the benefit of beneficial bacteria, enzymes, and proteins. Milk contains a constellation of anti-microbial activities including lactoferrin, immunoglobulins (antibodies), lactoperoxidase, and lysozyme. Once soured by harmless or benign acid-loving bacteria, as during the making of yogurt or cottage cheese, or by ordinary room-temperature souring (clabbering) when exposed to air, the acidity goes to work on any harmful bacteria, which may have entered the milk during standing, knocking them out rapidly. During the aging of hard cheese they are also destroyed.

Homogenization

As pasteurization increasingly made possible the resurrection of milk otherwise doomed, stopping bacterial growth in its tracks, a new problem evolved. Dairy economics dictated that the farmer, no longer in a position to make direct sales and stick the money in his pocket, must tool up for mass production and make it on volume. Formerly farmers, aided by wives and children, milked as many as sixteen cows twice a day by hand. Now ten times that many, nay, hundreds, can be done in the same time by machine. Unlike with hand milking, the machine does not know enough to stop when the udder is empty; overmilking predisposes to mastitis, an infectious inflammation of mammary tissue. With the best of intentions and constant rigorous preventive measures, it is nevertheless extremely difficult to completely prevent mastitis when cows are milked by machine. Mastitic milk is clotted and stringy and may contain antibiotics used to treat the infection. It is supposed to be thrown away, but sometimes it isn't. When cows are marching through the milking parlor ten at a time, the herdsman may overlook the warning tags meant to alert him to discard the milk from that cow. The mastitic milk used to be given to pigs. Now departments of health in many areas have banned the keeping of pigs on the same farm with cows. Without pigs, the milk must be dumped. The people who inspect dairies are not well-paid public servants. Sometimes, if farmers are generous, inspectors may forget to report a high leukocyte count in the milk.

Leukocytes, or white blood cells, are part of the body's defense system against infection. They are sacrificed in huge numbers "on site" when there is an infection. Their presence in milk is thought to cause not a health problem, but rather an aesthetic one. As described in chapter 1, after long standing, leukocytes settle down to the bottom of the milk and form a sludge that consumers not surprisingly find objectionable. Prior to the distribution of pasteurized milk in sterile containers, this was not an issue because milk went sour long before sediment developed. Moreover, the hand-milked product of a small dairy was unlikely to be mastitic. Homogenization solves the sludge problem by evenly distributing throughout the milk all the butterfat, leukocytes and other somatic cells, bacteria, and anything else. Although presented to the consumer as an improvement over old-fashioned methods because the milk is more uniform and seems creamier, homogenization, like pasteurization, is done to create an acceptable product using milk that would otherwise be rejected as undrinkable.

Homogenization has other pitfalls as well. The high pressure treatment breaks apart the fat globules exposing them to oxidation. Furthermore, the unnatural viscosity of homogenized milk leads many people to reject milk just because of the way it feels in their mouth.

Despite these objections, compared to juice, soft drinks, and soy imitations, pasteurized homogenized milk is a nutritional bonanza, especially if you can find nonhomogenized milk. And if fresh raw milk is available, there is just no comparison.

Fat

I do not encourage the use of low-fat milk. If weight control is the objective, carbohydrate restriction is more effective. Cream is extremely valuable nutritionally. It contains all the vitamin A or beta-carotene in milk. Natural home-produced milk, particularly when the cow is on grass, is also an excellent source of long-chain omega-3 fatty acids (specifically eicosapentaenoic acid). These fats are vitally important to health and are in short supply in most diets. They are essential to the integrity of the central nervous system and to normal hormone function. Cream also contains conjugated linoleic acid (CLA), a fatty acid with significant anticancer properties. CLA is found primarily in the fat of ruminants, including both meat and dairy products, and in particularly high levels in those ruminants that are grass-fed.

The fat in milk is valuable for another reason. In a country obsessed with weight control, it is easily forgotten that our first and most indispensable dietary requirement is for plain old calories. After that comes protein. In all babies and most of the ill or elderly, it is a physical impossibility to meet caloric requirements with carbohydrates and still have enough appetite to meet the protein requirement. The mere act of chewing can be fatiguing to those whose teeth haven't all come in yet or have been carried off by age. On a diet of milk alone, as in infancy, milkfat (butterfat) efficiently meets the caloric requirement. With skim milk the intractable caloric requirement will be partially met by lactose (milk sugar, a carbohydrate) and then fulfilled by protein; the protein will be broken down for its calories and consequently will not be available for growth and body building. Whole milk, on the other hand, contains sufficient fat to efficiently meet the caloric requirement of the nursling, thus sparing protein for its role as body builder.

The vulnerability of babies to a low-fat diet is extreme. In a less dramatic fashion all of us, but especially those with special needs, face the same situation. Those of us with freedom to visit the kitchen and the supermarket rarely deny our bodies the fat for which it clamors. Let us be sure we eat and serve our families fat from natural sources such as cream and butter.

Another fat factor that is almost universally overlooked is the role of fat in calcium absorption. In the absence of fat, calcium cannot be absorbed.

Taking these facts into account, in my opinion low-fat and skim (fat-free) milk come close to being worthless. Well, not quite worthless; diluted 3-to-1 they make fantastic fertilizer. Your rose bushes will salute you.

Lactose Intolerance

One reason many adults don't get along well with milk is due to a lack of the enzyme lactase, which is necessary for its proper digestion. It has been seriously suggested that this is proof that nature never intended humans to consume milk after infancy.

Nutrient requirements at the level of cell metabolism are virtually the same for all living things, and cells are not known to make inquiries into the origin of the molecules they receive. In all traditional cultures, past or present, people ate whatever was at hand where they lived. In some instances the menu was mostly fatty meat, in some instances mostly fruits or vegetables; in no case has it been devoid of animal protein. The keeping of animals for milk was common practice north and south of the equator thousands of years before the beginning of arable farming (grain). As a result, many breeds of animals have a symbiotic or mutual relationship with humans. This is because they can eat tough, inedible (to us) vegetation and convert it into milk and meat. We in turn smooth the path of life for them. Goats and cattle are at least as important in human history as dogs and horses. To assert that the nutritious fluid known as milk, the basic ingredient of life-sustaining foods used around the world, is somehow an inappropriate food not "meant" to be eaten is a shallow view of humans and one that ignores universal reality.

The absence of lactase in adults is a phenomenon primarily associated with people whose ancestors lived in warm climates, such as those of Africa and South Asia. People from northern populations usually remain equipped with this enzyme throughout life. People in warm climates kept (and still keep) many breeds of animals for milk. Before refrigeration, milk in warm climates was commonly made into fermented milk products using a controlled population of benign bacteria. When milk is fermented, the consumer has no need to produce lactase, because the bacteria do the job of splitting lactose, or milk sugar, the culprit that causes gas. Consequently there was no need for these populations to retain an adaptation for splitting lactose. Fermented or cultured milk products are properly highly valued for nutrition and flavor.

People who find fresh milk troublesome should use these cultured milk products, which include yogurt, kefir, buttermilk, and cheese. The great majority of lactose-intolerant people have no difficulty digesting raw milk because, cultured or not, it contains plenty of the appropriate beneficial bacteria.

Calcium

No other food commonly available in the American diet adequately takes the place of milk as a source of calcium. Vegetable and cereal sources are simply not reliable in providing calcium, because the calcium they contain is chelated (combined into indigestible form) and passes on through. Because of this fact, charts showing the calcium content of plant foods are essentially meaningless. Charts report the presence of the mineral, not its availability. Such charts have encouraged many people to trust, for instance, broccoli as a calcium source. No vegetables readily give up their calcium, and many actually grab additional calcium from the human gut. Further, fat is necessary for the uptake of calcium. Fat occurs naturally in whole milk but is often absent from vegetable and cereal preparations. The calcium in *whole milk* is readily absorbed, having been put there expressly to build bone. Yes, in the case of cow's milk it was put there to build calf bone. But any milk will do a pretty fair job of building bone for any other species. The similarities in milk from various species are more significant than the differences. Calcium is best absorbed from whole raw milk.

The diet of every successful human group has included some generous source of available calcium: for the Japanese, it was small fish eaten whole; for the Eskimo, little birds pickled and eaten whole and the stomach contents of large fish-eating animals; for the Indians of central California, flour made of dried grasshoppers. Persons with allergies to milk for whom such choices are unavailable or unappetizing need to find some other calcium source. Bone broth, especially if made with something acidic like wine, which dissolves calcium, is a good choice. And persons who tend not to tolerate milk often find that they do not have such difficulty with raw milk.

Food Allergies

Food allergies are caused by microscopically small particles of incompletely digested proteins passing through the gut wall and entering the bloodstream, where they trigger an allergic response in a previously sensitized individual.

The response can take many forms, and its intensity is unpredictable, varying with the nutritional state of the victim and often worsening under conditions of stress. In the case of milk, the nutrients of which are difficult to replace with other foods, every effort should be made to overcome the allergy. Often the switch to untreated (raw) milk is effective; if not, then buttermilk, yogurt, or other cultured dairy products can often be tolerated.

Some allergy sufferers are so exquisitely sensitive that components in foods such as soy (a common allergen) in the diet of the cow will affect them by passing through into the milk. By controlling the diet of your cow, you can provide such individuals with "designer" milk. Many people who believe themselves to be allergic to milk or lactose intolerant actually are responding to accidental inclusions in commercial milk. Besides items in dairy feed, iodophors used to clean milk pipelines or chemicals leached from plastic milk containers may be the real culprits behind dairy intolerance. (You can't escape plastic by switching to cardboard containers; they are lined with it.)

Keith Woodford, author of *Devil in the Milk: Illness, Health, and the Politics of A1 and A2 Milk*, makes a persuasive case that a recently identified casein protein in the milk of many modern cows is the cause of milk intolerance (see page 125 for more discussion). Many disorders attributed to allergies may be a reaction to this casein fragment, and it is common in commercial milk. So don't give up on milk forever just because you or a member of your family has had trouble with it in the past. You can test yourself (or your family member) for sensitivity to this particular protein, and you can test any cow you're thinking of raising to make sure it does not produce milk with this casein.

Historic accounts make it clear that there have always been people with allergies. However, the presence now in our population of large numbers of people with cow's milk and other allergies is thought to be due, at least in part, to the practice of formula feeding of infants. This is a modern phenomenon. Formerly almost all babies were fed mother's milk. When this is done, allergies are rare.

The Question of Atherosclerosis

What about the belief that milk drinking predisposes to atherosclerosis? This artery-clogging disease was little known before the twentieth century; its incidence has risen in concert with increased consumption of sugar and vegetable oils, progressive depletion of vitamin E and omega-3 fatty acids in foods, and the creation of artificial fats by hydrogenation (trans fats). The per capita consumption of milk, butterfat, and animal fat from all sources dropped steadily throughout this period. This fact can be confirmed by consulting the clear graphs printed in the annual yearbooks published by the U.S. Department of Agriculture. It is thus highly illogical to place the blame for heart disease on milk or any other animal product. An increasing number of researchers and clinicians recognize this. The role of sugar as villain now has many adherents. Among the first to make the case against sugar was the noted British scientist John Yudkin, who repeatedly complained that sugar interests blocked publication of his remarks. Strong support for dairy products also comes from important sources in the United States. Dr. Roger Williams, discoverer of the vitamin pantothenic acid, considered that avoidance of dairy products as a defense against heart disease was foolish indeed. In his book *Nutrition Against Disease*, he presented an extensive review and analysis of current theory on the causes of heart disease. This is no simple matter, since many nutrients are involved. A few factors emerge clearly, and perhaps of foremost interest to the consumer of dairy products is that despite enormous publicity regarding the avoidance of animal fat and concern over exogenous cholesterol (that which is obtained from food), heart disease has not been controlled by the avoidance of dairy foods.

Vegetable Oils

The substitution of vegetable oils for animal fats brings its own hazards while being useless in the control of heart disease. According to Williams, except in the presence of adequate vitamin B_6, vegetable oils have been shown to actually predispose to atherosclerotic lesions. Clarified and deodorized oils, as they are commonly sold, are heated to a high temperature for several hours and are probably dangerous to use.

Nutrition authorities have ceased to recommend most oils due to their propensity to develop free radicals, long recognized as carcinogens. When hardened into margarine or shortening, oils are altered into an unnatural form of saturated fat. When only partially hardened, they form trans fats. Trans fats have been proven to interfere with important metabolic functions and are harmful. These chemically altered fats are preferred by processors because at room temperature they are not runny and greasy like oil, when cold they are not hard like real butter, and they are cheap to manufacture.

Research on cholesterol and animal fat as factors in heart disease has now been abandoned as a dry well. Reluctant to admit error, around 1990 diet gurus switched to counseling against *all* fats, thus avoiding uncomfortable discussion of the rapidly emerging evidence against fats of vegetable origin.

Williams cites numerous studies of human groups subsisting on diets extremely high in animal fats (including butterfat) and protein. These groups are virtually free of heart disease. They include the inhabitants of the Loetschental Valley in the Valaisian Alps of Switzerland, the primitive Eskimo, and the African herdsman tribes (Masai, Somali, Samburu), who live almost entirely on milk and meat.

The reader interested in further details and an outstanding bibliography is urged to obtain a copy of *Nutrition Against Disease*. Williams was a scientist with a lifetime of research in nutrition. His books and work have been the inspiration for many other scientists, including Linus Pauling. He is but one of many who never subscribed to the theory that dietary cholesterol is a risk factor for heart disease.

Applying Logic to Nutritional Claims

Except for olive oil, I rarely use vegetable oils. The use of vegetable oils in the quantities found in modern foods is unprecedented in human history. Their use constitutes a major dietary experiment carried out on a public now badly confused by false advertising. A few facts are indisputable:

1. The consumption of sugar increased steadily throughout the twentieth century and remains at an all-time high in the twenty-first century. It is augmented by the use of high-fructose corn syrup.
2. Our consumption of fats of vegetable origin is the most rapid change in human diet ever known. The average intake now stands at about one cup per day for each American. The processing received by milk is gnat's breath compared to what vegetable oils endure, yet oils and oil products are permitted to be labeled "all natural."
3. The incidence of heart disease and cancer has risen in a matching curve with the use of vegetable oils.
4. The consumption of fats of animal origin has fallen steadily since about 1890. Use of animal fats follows an inverse curve to the incidence of heart disease and cancer; animal fats down, disease up.
5. The argument that we are now seeing more degenerative disease because of increased longevity, rather than dietary and other changes, doesn't hold up. This belief is due to widespread misunderstanding of where those numbers come from. Life expectancy is normally figured

based on the total population of people born. In previous times, infant and childhood mortality rates were high, and these numbers skewed life expectancy figures downward. Yet a person who survived childhood was likely to live a good long while. The twentieth century produced a notable drop in infant mortality. When you correct life expectancy figures for this, using only deaths of people past early childhood, at most we see a couple of years, if any, added to modern life expectancy rates. Think of it this way: If I turn in ten spelling tests to my third-grade teacher and get 100 percent on each, my average score is 100. But if she refuses to count five of them because she can't read my writing – in other words, five of them die in infancy – my average score drops to 50. I'm not a great hand at math (or spelling), but I'm astonished at how many people are confused by this. Last Christmas our vicar declared that the Virgin Mary was already middle-aged, her life half over, when she conceived at age fourteen because she lived in an era when the life expectancy was thirty years.

6. The incidence of obesity bears virtually no relationship to the consumption of animal fats. It is, in fact, related to the consumption of carbohydrates. This fact was never lost upon farmers, who have always used grains to fatten animals. The fact that obesity results from overconsumption of carbohydrates is increasingly being recognized by the general public as a result of several popular books by nutrition writers with grounding in biochemistry and endocrinology.

The campaign against dairy foods and other animal products is motivated by something other than fact.

The remaining critics of fresh dairy products are certain to fade before the mounting evidence, not just of their safety, but also of their positive importance as a source of available calcium, phosphorus, magnesium, long-chain fatty acids, essential amino acids, anti-stress factors and enzymes, fat-soluble vitamins, and undegraded (undamaged) protein. And let's not overlook flavor.

Babies Require Cholesterol, and Some Other Interesting Facts

Babies and children require cholesterol. Without it they don't thrive and their central nervous system is imperiled. If fed a low-cholesterol diet, their bodies respond by rapidly generating cholesterol. Animal studies have shown that this results in the body's cholesterol handling system becoming permanently deranged. Studies of older people have revealed greater longevity among those who did not avoid cholesterol; avoiders had a disproportionately high incidence of cancer.

The naturally occurring fatty acid CLA (conjugated linoleic acid) found in the milk and fat of grazing ruminants, and to a limited extent in products from confined cows, has been shown to act as an anticarcinogen in animal studies. It is speculated that CLA may be the agent responsible for the lower incidence of breast cancer in women who are habitual milk drinkers.

Why Doesn't the Dairy Industry Tell Us Any of This?

One may well ask why the National Dairy Council (a promotional organization) or the Dairy Farmers of America (a marketing group) does not arm itself with some of this evidence and fight back against the processed food industry that has flooded the market with fake foods under the specious anticholesterol banner. This question has puzzled me for years. Some people apparently assume that dairymen have fought back, for in health magazines and trendy books on diet the authors often encourage readers to forget the "dire threats by the dairy industry." I've been waiting for just such a show of indignation by dairymen for years but have seen only vapid pictures of apple-cheeked children or celebrities drinking skim milk in their generic advertisements. Dairy farming has been a captive industry since the 1930s, when the first great wave of regulations and subsidies began. The processors soon had dairy farmers in their grip. Now major processors of milk products also market imitation dairy foods. These fake products are advertised right on their butter wrappers.

Processors lack motivation to provide the facts on the poor nutritional value of low-fat (skim) milk because the absent cream is directed to the manufacture of ice cream, a value-added product. Casein (a milk protein extracted from Grade B milk), surplus Grade A milk, and whey solids are found in countless processed foods. Casein and whey have escaped their identity as dairy products and are now "commodities," thus freeing them for unrestricted use in processed foods. They are freed also from restrictions imposed on real dairy products. Unlike brand-name products, real foods such as milk, meat, fruits, and vegetables can only be advertised generically. The real money is in processed foods.

Dairy farmers themselves have virtually no influence on what is done with milk once it leaves the farm. They have no influence on claims made for or against milk. They have never successfully organized. Many seem oblivious to the larger issues. They do their best in their fourteen-hour day.

Jersey and Holstein Milk Compared

Jersey milk is creamier than that of most other cows; it is also better endowed with every ingredient except water. Ah, yes, *water*. I once read that when state restrictions in the early part of the twentieth century made it illegal for farmers to water down their milk, they reacted by breeding a cow that would do it for them. The cow they developed was the Holstein-Friesian, the familiar black and white. Her milk is 3.0 to 3.9 percent butterfat (now referred to as milkfat), close to the legal minimum for whole milk, compared with 6.0 percent for many Jerseys. The ratios for protein, calcium, phosphorus, magnesium, and all the other minerals track fairly closely with fat. This means that if for every two cups of Jersey milk you were to add one cup of water, you'd get Holstein milk. Without partial skimming, Jersey milk is actually too creamy for the taste of some people. So if you have a Jersey cow, you'll have plenty of cream left over for making butter.

I have used the Jersey here for purposes of comparison. The Guernsey, Devon, Dexter, and a number of other breeds and crosses approximate and sometimes exceed the Jersey in milk solids.

Value of Raw (Untreated) Milk

Once your household becomes accustomed to the fresh dairy products you will be serving them, no further recommendation than the delicious flavors will be necessary. But there are more advantages to raw milk. The bone-building nutrients in raw (unheated) milk are more fully absorbed in the gut. Several very important nutrients, including pyridoxine (vitamin B_6) and vitamin C, are damaged or destroyed by heating. Fresh milk is a significant source of vitamin C, whereas store-bought milk has virtually none. Riboflavin, a B vitamin essential to growth and eyesight, is found in few foods besides milk. It is extremely fugitive, being quickly destroyed by light and heat, and so it is present only at greatly reduced levels in commercial supplies stored in lighted cases.

Heat changes protein in a way that causes it to be less valuable to the human body. Several important amino acids, including tryptophan, lysine, and methionine, are progressively broken down when milk is heated.

Even though pasteurization does not cause complete loss of nutrients, why lose any if not necessary? Fresh milk as it comes from the cow has vitality. It contains useful enzymes. One of these enzymes, called the Wulzen antistiffness factor, is entirely destroyed by pasteurization. When I returned last spring to my farm after a winter away I was stiff in every joint. After three months on fresh milk I noted scarcely a twinge. I'm inclined to give much of the credit to Wulzen. Following heat treatment (pasteurization), milk contains few of these active principles, even when organically produced.

If the cow is on good pasture, milk also contains an array of long-chain omega-3 fatty acids, those essential fats otherwise available in high-priced capsules. These fats, such as eicosapentaenoic acid (EPA), are also available from deep-sea fish such as salmon. Omega-3 fatty acids are believed to be capable of dramatically lowering the risk of heart disease. Yet their absence from dairy products, beef, and eggs is often cited as a reason to avoid these foods. Nevertheless, like CLA, omega-3s are absent from animals raised primarily on grains; they are *not* absent from animals raised on a natural outdoor diet. (Some EPA survives the heating of milk, meat, and eggs from grazing animals but it is unlikely that any survives homogenization.) Critics of animal foods including dairy products would be better advised to campaign for more natural feeding practices, not avoidance of vital foods.

The active principles in raw foods such as milk are mostly destroyed following heat treatment. So it is important that a significant part of the diet of both people and animals be raw foods. Ideally, every meal should include something raw. This can be difficult to achieve — unless you have a cow. Experiments with animals observed over several generations have demonstrated the impossibility of even keeping the subjects alive without the inclusion of raw food in the diet. Such experiments have had to be terminated because of reproductive failure or cannibalism.

There are few recent studies on the comparative performance in the diet of heated (pasteurized) versus unheated (raw) milk. The bright graduate student interested in carving a place for him- or herself among the scientific archives prefers to write his or her thesis on something that has assured support. And the choice of a research subject is guided by the professor under whom the student is working. The choice will usually be an ancillary aspect of the professor's own work, and something for which grant money is available. There is hardly any grant money available in departments of dairy industry for work on the classic dairy products, let alone raw milk. A lot of money is available for food technology in its various aspects, such as the control of off flavors of food sold in aerosol cans. If we require further proof of the virtues of untreated milk, we shall probably have to rely upon fallout from research in other areas of study. For instance, there is new scientific awareness of the significance of enzymes in food, especially in conditions of stress. Fresh raw milk is a rich enzyme source.

Probiotics may be another field in which research money is to be found. Support for the immune system is provided by benign bacteria found in raw milk, particularly milk soon after it comes from the cow. Some of these friendly bacteria, known as probiotics, have remarkable disease-fighting properties. They line the gut wall and induce the production of cytokines such as gamma interferon, which has potent defense capabilities. Research interest in probiotics is intense. To date it appears to be focused on isolated strains of lactobacillus, which are added to juice or milk formulas intended to provide protection against yeast infections. Interest in raw milk itself has not so far been a theme of this research, although hatcheries are now misting baby chicks with whey, which has been found to protect them against coccidiosis.

The active factors in raw milk no doubt help account for the "bloom" that distinguishes the cow-fed calf from the artificially fed calf, and which is also to be seen in the complexions of the fresh-milk family. This bloom is the glow of health.

I once knew a family with eleven children that lived eighty years ago in an isolated area of Maine long before convenience foods. They seemed to live most of the year on nothing but home-baked beans and the milk from their cow. They were all magnificently healthy and had excellent teeth. I wonder if their descendants look as good.

Good Bone Structure Is No Accident

Within the strident propaganda against animal fat and other animal products, all of which focuses on degenerative disease, the importance of growth and development has been ignored. Evaluation of bone structure and muscular development among young people seems a forgotten art. No amount of tooth brushing will straighten children's teeth, yet the teeth of children raised on the milk from their family cow are always straight. I'm making this rather sweeping statement because I have never run into any exceptions. I have had letters from readers describing the freedom from decay their family enjoys now that they keep a cow. And I have raised a large family with straight teeth, legs, and shoulders.

Environmental Impact of the Dairy Cow

The belief that cattle have an adverse environmental impact and compete with humanity for food – first introduced by Frances Moore Lappé in her book *Diet for a Small Planet* – is without foundation. Nonetheless, this belief has now assumed almost mythic status and has resulted in compromised health among caring people attempting to do their part for the planet by avoiding beef and dairy products. Chapter 9, "Feeding Your Cow," should make clear the irrational basis of this belief.

Lappé also set forth the attractive but equally unsubstantiated theory that combinations of plant foods called "complementary proteins" are the nutritional equivalent of animal products. My search of the literature and queries to researchers have so far failed to discover any valid source for this attractive but overly facile theory. The bibliography in *Diet for a Small Planet* lists two books as research sources for the complementary protein concept. Neither reference actually contains the least corroboration for the idea. Yet the presumed adequacy of a plant-based diet remains the ultimate fallback position for every anti-animal-product argument.

Feedlot fattening of cattle, a man-made evil, has been used to condemn the entire bovine species. It is not the fault of cattle if they are bunched together, they are fed a lot of grain, and their manure is allowed to contaminate the environment. Our environment and family health are both best served by small farms or individually owned animals. The efficiency of this model withstands any analysis that compares energy in versus energy out.

There is a Chinese proverb that says:

> *I hear and I forget.*
> *I see and I remember.*
> *I do and I understand.*

If you still have doubting friends, invite them over to share a day with you and your family cow. They'll begin to understand.

CHAPTER 3

Milking Your Cow

John Seymour, in his superb book *Self-Sufficiency*, devoted three chapters to the cow and her products, yet to instructions for milking only one sentence: "Now sit down and milk your cow." Milking, like riding a bicycle or making bread, does have to be learned by doing, but the suggestions in this chapter may help.

A cow is customarily milked from her right, and if your cow was previously owned, this is what she will be accustomed to. The reason has to do with the slightly greater reach and strength offered by your dominant arm – which for most people is the right one. If you are left-handed, there is no reason you can't switch sides, provided your cow is amenable.

You will need something to sit on, and it must be very low and suited to your height. It should be light and easily moved in case the cow shifts her position. The classic milking stool has three legs, making it easy to rock forward as needed. It often has a handle like a pan, the better to grab it when the cow shifts position. It is the rare cow that stands motionless throughout milking.

Once a cow has come in to be milked three or four times, you will need merely to open gates and doors and she will walk to her position. Have everything ready beforehand, including closing any gates or doors you do not wish her to enter; she is sure to notice if you've left something open. It is customary to give the cow her grain at this time, and if she knows it is waiting for her, you may be sure she will come in readily. If you don't feed grain, try cut-up apples or carrots. Don't ask her to walk up any steep ramp or across slippery spots. She will do it but will soon injure herself, sometimes very badly. This is the voice of experience. Preferably do not make her wait in muck or walk through it on the way in. You want her feet to be as dry and clean as possible.

If flies are bad, it helps to set up a system to brush them off as she comes in. Otherwise they will ride in on her back. You can hang up a curtain of

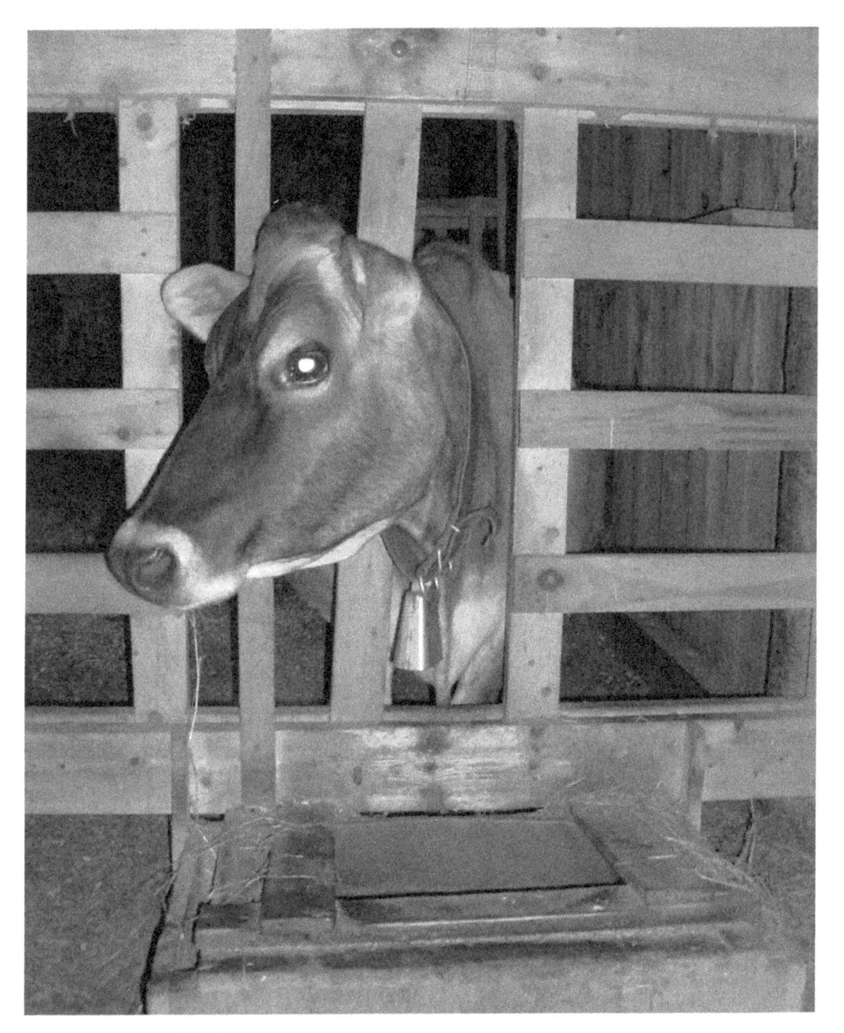

Photograph courtesy of Max Luick

bags or string or just sweep off the flies with your extended arms so they jump up in the air. Flies do not like to enter the dark, so if you dim the inside lights until the cow walks in, you can have fly-free milking.

You will find it a great deal easier to milk if you have a perfectly flat floor that is not slippery. A cow must have some grip for her feet. Cement that has been scored lightly before drying will do, but it is not ideal because it is hard to keep clean. And new cement takes months to quit leaching harsh chemicals. Rough wood works fine. It makes comfortable standing. Keep a bucket of lime handy and throw some down anywhere your cow might slip.

A locking stanchion or headgate works better than a simple tie-up to keep a cow in place for milking. A tie-up permits a cow to move away from you or toward you, making it necessary for you to shift your stool backward and forward. If your cow is not accustomed to being in a stanchion, tact is called for. Teach her that it is not a cow trap but is instead a lovely spot for a snack. Put treats in the feed pan and let her find her way into it a few times without locking the gate. When you finally lock it, stand by with more treats and some sweet talk. For a stoic older cow going to all this trouble may be unnecessary, but I have known a cow to become frantic the first time she was locked in. If this happens, she will be reluctant to stick her head in again.

If you have not built a stanchion, fasten her with a short rope or chain, using an easy clip.

Unless she is unaccustomed to the concept of being milked or to your family, your cow will probably stand still until her grain is gone. This is a grace period that you don't want to squander. That's why you want everything standing ready.

What to Have Ready

First, brush the cow's back, sides, and belly with a stiff brush to remove any debris.

Have ready a bucket with several quarts of quite warm water to which you have added a couple of ounces of vinegar. Alternatively, use a squirt bottle containing vinegar and a little liquid soap and squirt this on a cloth to wet it. This mixture will keep her udder and your hands free of chapping as well as anything else I have found, and so far as I can tell, it is a successful germicide. I keep a bag of clean white cloths beside me and use as many as needed, never double-dipping a cloth that has touched the cow. The hot water stays perfectly clean. Give her whole udder a good wash with special attention to the teat ends. You'll need a good light for this. She won't mind hard scrubbing. A final rinse with plain water is not necessary. Dry her with another towel. Get her good and dry so there is no chance of drips into the milk pail. By this time milk will usually be streaming, as the warm water and towels stimulate letdown.

If not actually dripping, her teats should be fully puffed up, not withered looking. Filled teats are a sign she has fully let down and is ready to be milked. This is especially important if you are machine milking, as applying the vacuum system to a dry teat may be injurious.

How to Milk

Take a couple of squirts from each teat, avoiding your milk pail. This gets rid of the waxy plug sealing the teat; also, the first squirts have a higher bacterial count.

Position yourself and the pail and grab a teat in each hand. Do not be tentative about this. Cows hate a tickly approach. Most people put their head or shoulder into the cow's side. Wrap your thumb and forefinger around the teat as high up as possible, squeeze the teat off so milk does not backflow, and, with all your fingers firmly wrapped around the teat, *pull down*. You don't need to pull down hard. Alternate squeezes, left, right, left, right, in a steady rhythm. If your cow is newly freshened or is a heifer, the teats may be small or partially effaced. Time will eventually cure this, but in the meantime, you may have to milk using only two or three fingers, at least until some of the milk is out. If, on the other hand, your cow is old and has long teats, position your hand low enough on the teat so that you are not creating a ballooning effect at the teat end. Ballooning will damage the teat end and predispose to mastitis. Also the cow doesn't like it and will switch her tail. Milking success depends upon coordination and persistence, not strength. Just because you have trouble opening ketchup bottles doesn't mean you can't be a good milker.

As you alternate squeezes and before each pull, push your fisted hand up into her udder. This simulates the bunting of the calf and maintains

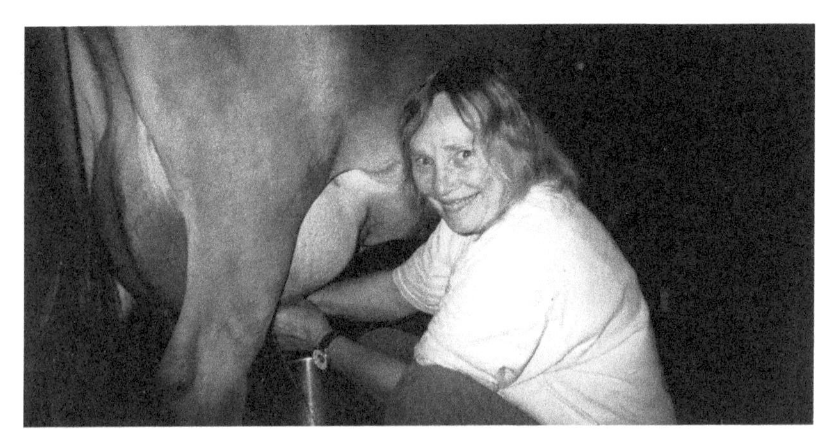

The author milking Helen.
Photograph courtesy of Joann S. Grohman

letdown. Have you driven through farming country and seen those wooden whirligig lawn ornaments of a little farmer milking his cow, his arms flying up and down? Soon your arms, too, will be flying up and down.

Aim for a good, steady rhythm with as few breaks as possible, and aim for speed. When you are able to finish with a strong layer of foam on the milk, consider yourself a pro. I usually milk the back teats about halfway out first, move to the front, then return to the back and finally the front again. Pause occasionally to bounce the cow's udder around a bit to shake loose the milk. Finish by doing each teat singly while you knead the quarter of udder above it to be sure you've got all the milk. Aim to get all the milk and ask for more until you're down to tiny squirts and all four quarters feel limp.

There are two reasons that it is important to get all the milk. Milk left behind is an invitation to mastitis. And milk left behind is a message to the udder: don't bother making so much milk next time; nobody needed it. Thorough milking is of key importance to maintaining high production and udder health.

Milking the Fresh Cow

When a cow first freshens (calves), her udder may be alarmingly swollen and her teats nearly effaced. Get at least a little colostrum out of each quarter to make sure it is flowing. Most people agree that it is best not to take it all for a couple of days – as if you could. There may be very little teat to take hold of. If her udder is alarmingly swollen, try to determine whether she has mastitis or edema. Edema is fluid seeped out of the blood supply to the udder. Push your finger into it. If a dent remains, it is edema. Help the swelling to leave by massaging upward toward the cow's belly. A lotion such as Uddermint is helpful. Mastitis will also cause swelling, but unlike edema it will be localized to one quarter or part of a quarter and will feel hot. If you believe the swelling to be mastitis, that quarter does need to be milked out as completely as you are able. If the case of mastitis is mild and the cow has not been medicated, the milk need not necessarily be discarded. Breastfeeding mothers nurse their babies right through mastitis. But if in doubt, feed the milk to your rose bushes.

Exercise will help in both prevention and treatment of edema. Get your cow to walk around.

How much to milk out after freshening and when to milk out completely is a difficult judgment call. And with any cow that has just calved, watch for milk fever (see chapter 15).

My Aching Back

When you first learn to milk, your back will ache and your hands will be in pain and there will be no foam on the milk. It's like getting into any new sport. There is going to be pain until your body rises to the challenge, which it will if you persevere. Here are some things that will help.

Take extra vitamins C, D, and E. This will aid in preventing and recovering from sore muscles. If you note a tendency to cramp, be especially sure to take these vitamins and also to drink plenty of milk yourself for the calcium, a great aid against cramping.

Don't forget to breathe. I find it's easy in a touchy situation, when I've finally got her standing just right, almost to forget to breathe. But milking is an aerobic sport. If some part of you is feeling numb, don't assume it is repetitive motion disease or Raynaud's syndrome; just take some nice deep breaths and feed your fingers and calf muscles some more oxygen.

Wear loose clothing and very comfortable shoes. Oxygen can't reach your toes if your blood can't get past your tight waistband and constricted instep.

Experiment with the height of your milking stool. It needs to be exactly suited to you. If there is more than one milker in the family, each will need a customized stool.

While milkers don't need to be exceptionally strong (many children are good milkers), a degree of strength is required. I often find it helpful to arrange my knees so I can brace my elbows on them as I milk. This is especially helpful during the first ten minutes of milking when the udder is full and heavy.

Cows are creatures of habit. This makes handling them easy once they understand what is wanted. They like to do the same thing every time. But some habits can be irritating. If your cow gets in the habit of dropping manure or urinating three-quarters of the way through every milking, here is a trick that has worked well for me. You can often tell when she is about to drop manure by the raising of her tail. Arise quickly, moving the bucket aside, and grab a large shovel that you have handy. Place the shovel against her thighs and catch the manure on the shovel. She will think it's pretty weird having a shovel against her thighs and not hearing the usual plop . . . I guess. In any case, this seems to take the fun out of it for her without doing her any harm, and in a day or two she'll quit the habit. Curing urination is even easier because the necessary change of stance gives you an extra second of warning. Have handy an empty five-gallon plastic bucket and simply catch the urine in it. This makes such a noisy splash that she will probably stop halfway through. Urine is superb high-nitrogen fertilizer, so don't waste it. Dilute it three-to-one with water and use it on any plants you wish to favor.

I also tell her what I think.

Kicking

Even the mildest cow will occasionally kick the bucket. This is so aggravating that I have a theory that the phrase "kick the bucket" as a euphemism for death arose from occasions throughout the ages when the milker stood up and whammed the cow over the head with the nearest heavy object, with fatal results. ("Yeah, she kicked the bucket.") Better to plan ahead for it, reminding yourself it had to happen sometime, yes I'm angry and I'm telling the cow about it, but I'm not going to cry over spilt milk. I'm making my point to her effectively, I'm analyzing why she did it, and I'm instituting preventive measures.

Sometimes it's obvious why she kicked. A kitten used her leg for a scratching post. A chicken flew down from the rafters and frightened her. Your neighbor roared up on his motorcycle. The flies and heat are getting to her. You still need to express your disapproval just to reinforce her future self-control. I yell and grab her leg and plant it firmly down and finish off with a pithy lecture.

Sometimes the kick seems unprompted by anything obvious; she just seems to be in a bad mood. The commonest causes for this are estrus (heat) or being milked a lot later than she likes. Cows are like babies; you may think you aren't going to be ruled by their fussing about their schedule being flouted, but after a while they train you. It just gets a lot easier to do it their way. Cows don't "have" to be milked at the same time every day. They just like it a lot better, and pretty soon you will too.

Cows develop strong attachments, and some are more conservative than others about accepting a substitute milker. Kicking may ensue.

A first-calf heifer may do a lot of kicking before she accepts her appointed role. She often will even kick her calf. Be sure she is firmly tied or in her stanchion. Have somebody stand behind her and hold her tail up very high. Not to break it, just to immobilize her. Something about this tail raising enervates her leg movement, and cows show no signs that it hurts. When she no longer kicks, you will know the tail is high enough.

Alternatively, a rope trick that prevents kicking involves making a cinch around a cow's belly just in front of her udder. A helper has to hold the end of this noose unless you are better at knots than I am. Tighten it until she stops kicking. It must be released after three or at the most four minutes, as it stops blood flow to her udder and can even cause her to tip over if too tight. It is really very dangerous to leave any rope tied onto a cow because someday you will forget and let her out with the rope attached. The results can be disastrous. But as an anti-kicking measure this is one of the best. It has the advantage that once the cow has understood that the rope means "no kicking," with the rope loosely on the cow, the milker, without changing position, can give it a touch to remind her. Another advantage I believe is that the rope is an impersonal aid. Tail hoisting, which is also likely to be messy, is unmistakably something humans are *doing* to her.

There are anti-kicking devices designed to lock onto a cow's hocks and prevent kicking. Another type resembles a giant C-clamp and fits from in front of her near hind leg up and over her backbone. Once adjusted it goes on easily but firmly. All of these approaches can teach her you won't put up with kicking. Whatever method you use, avoid frightening her if you can, as fear is a hard habit to break.

Training the Determined Kicker

Lee Anne B., moderator of the wonderful Keeping a Family Cow online forum, has found a less invasive training trick that has been highly successful for her and others. She describes it thus:

> *Run a broom handle into the middle finger of an old winter glove, and duct-tape it on securely. Secure your recalcitrant cow in her stanchion or tie-up, and stand near her shoulder, out of kicking range. With the end of the broom handle in your hand, and the body of the handle near her belly, place the "glove on a stick" against her udder. She can kick all she likes without risk of injury to you, and no matter how much she kicks, the "glove on a stick" will not go away. If she kicks it out of your hand, just pick it up and put it right back. It may take two minutes, or it may take twenty, but eventually she will settle down and stop kicking. Praise her gently, stroke her shoulder, and, still holding the end of the stick, rub the glove all over her udder, teats, belly, and back legs. Use the glove to touch her calmly but purposefully*

everywhere you will touch her when you are milking. If she kicks, hold the glove in that spot until she stops, and then rub it over the spot that gave her offense until she accepts it calmly. Then take a few deep breaths, grab your stool and bucket, and get to milking!

The kick of a cow is never, so far as I know, fatal, and seldom even seriously dangerous, but it can certainly make milking impossible. Unlike a horse, a cow is not an athlete. Her kick is analogous to what a human can do without bending the knee. If truly angry or frightened, she can kick backward with her front leg using a short flipping motion.

She can also whip her tail around. This can be annoying. In desperation you can tie it down to her off hind leg. If you tie it to the wall instead, use a light string that will break away. Otherwise someday you'll forget to untie

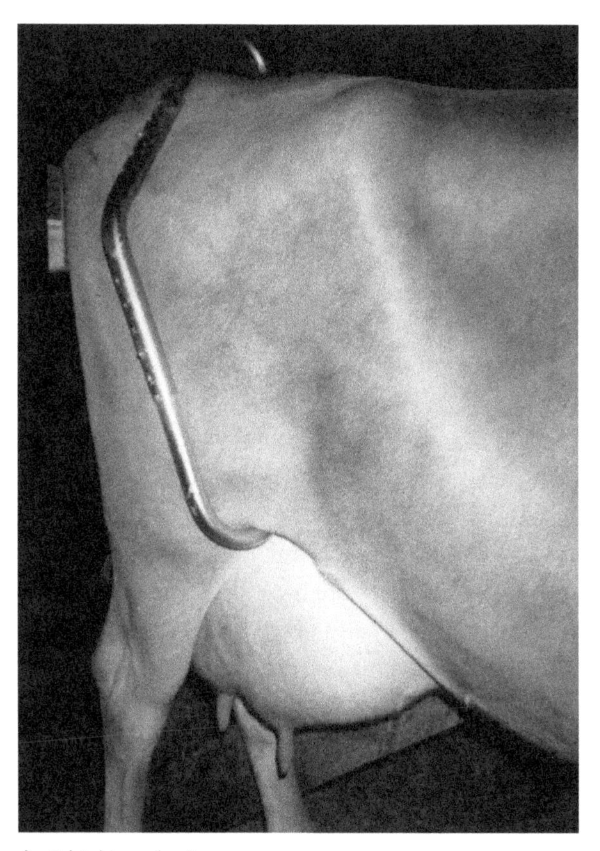

Anti-kicking device.
Photograph courtesy of Max Luick

it when you let her out and her tail will break when she walks away. Resist the temptation to cut off the tail hair (called her switch). When it reaches to the ground it is just long enough so she can flick flies off her withers. The hair grows back very slowly. It may take a year to grow out. If it is dirty, it is preferable to soak or comb it clean rather than cutting it. This may not seem important in winter, when the switch is most likely to get filthy. But if she can't switch flies to her satisfaction, next summer she will spend a lot of time hiding in the shade when she ought to be grazing.

Dairy cows have been bred for centuries not only for milk production but for good temperament. As with all creatures, there are different dispositions, and if you have ended up with a truly intractable cow, get rid of her. Most dairy cows, while kicking occasionally, are pretty cooperative and kick only when they think they have a good reason.

A Calm Environment

The way to make milking a pleasant experience for all concerned is to create a calm milking-parlor environment. You may then go for many weeks without incident. Here are some factors to bear in mind. You may get away with one or two deviations from this list, but here is what cows like best, as well as some situations they can be counted on to dislike.

Cows like quiet, contemplative music. They don't like raucous music and they don't much like newscasts either. Actual dairy farm studies have shown this clearly. A recent study showed a preference for classical over rock, but that is a mere average reflecting the fact that over a day's listening there is a little less cacophony in classical music. But I promise you they don't like Schoenberg any better than rock, and they hate string quartets; I suspect they are reminded of swarming bees. They especially like music from the *Hearts of Space* program or anything dreamy.

Cows don't like dogs. They are natural enemies. Exactly as with music a cow doesn't enjoy, you may not notice that she tenses up around a dog. A lot of cows averaged over many hours of listening were required for the discovery that rock music cut production by 4 to 6 percent, not a margin you would think to attribute to your music . . . or your dog. But when she is walking past the dog you will see the signs plainly enough, especially if her calf is anywhere in the picture.

Children who run or giggle are poorly tolerated by cows. Any but the quietest strangers will cause her to stop letting down. Most dairy farmers are not very hospitable about allowing strangers into the cow barn, whether it's milking time or not. The first comment a cow makes is usually to raise her tail and make a plop. A farmer prefers not to clean up behind the whole row.

There are two schools of thought on feeding a cow at milking time. It has traditionally been done because it is convenient and it keeps her quietly occupied until she has cleaned up her feed. This will occur long before most people are finished milking. She then is restless for several minutes while she double-checks around and under everything her tongue can reach in case there is a crumb of grain she missed. Once satisfied nothing was overlooked, she is likely to enter a quiet meditative state and you can finish milking peacefully.

If this isn't working for you, don't resort to more scoops of grain. The amount of grain a cow gets should be something you have decided on after reading chapter 9, "Feeding Your Cow." If you want her to keep eating for any of several good reasons, give her something that won't unbalance her diet. Some people slow down their cow by putting rocks in her feed pan. I myself may offer chopped dried alfalfa at this time. Most cows love it, and it is good for them.

Cows can be habituated to come in to be milked without providing any feed. Some people find the cow settles down more easily this way. Without the temptation of feed, unless the cow is already confined, a regular schedule is important so she shows up at the gate for you. If your cow is a heavy producer, she will look forward to the relief of being milked.

Lastly, and at the risk of forfeiting my reputation for scientific objectivity, I find that it helps to keep the cow calm and settled if, while milking, I avoid any worrisome or peevish thoughts and make a point of visualizing pleasant things. Cows are extremely sensitive. Janene R, long time moderator on the Keeping a Family Cow forum, puts it this way: "If you're already in a bad or a hurried mood, she'll know it and pick up on it and will be hesitant/expectant of bad vibes/mojo (for lack of a better term). You knew when your parents were upset with you and expected the worst. The cow can do the same."

When milking by hand, it takes me just about one hour to complete the entire milking chore without help, from assembling the equipment in the kitchen to returning with the milk and completing the washing-up. Time spent actually milking is dependent on current production but is close to twenty minutes.

Machine Milking

It is perfectly feasible to set up machine milking for just one cow. There is plenty of used equipment about. A milking machine is just a vacuum pump with a hose attached to the milking unit itself, which is a set of four rubber-lined teat cups, called the claw or cluster. The cups merge to deliver milk to a stainless-steel container. A unit called the pulsator regulates the vacuum in an on/off rhythm similar to the sucking of a calf or your hand milking. If you are a handy person more drawn toward technology than the aerobic yet meditative art of hand milking, there is no reason not to consider installing a milking machine. You will need to do some studying on your own. If possible, take the mystery out of it by visiting a small dairy or another cow owner and see one in use. There is also a trove of information online, which will acquaint you with the various models.

If you must buy sight unseen, it is best to order your machine from an established dealer who can offer guidance over the phone. This is especially important if you are buying used equipment. Milking machines hold few mysteries for me now, but lacking such help, I have spent many a morning in tears of frustration as I tried to solve what usually turned out to be a minor problem. I do urge you to buy a late-model pulsator. The early models work but require a lot of maintenance. Many helpful people can be found at the forums on the website for this book (http://www.keepingafamilycow.com).

One advantage to a machine is that you may find more people willing to learn how to use it, rather than learning the skill of hand milking, thus gaining yourself freedom. If there is someone willing to assume responsibility for the twice-daily machine wash-up, then the task of milking is cut to about seven minutes.

Disadvantages of the machine, apart from the initial cost and upkeep, are:

- The milk rarely tastes as good once it has been through a machine. Rarely does it keep as well. Nor does the cream rise as thickly.
- There is no saving in time when milking only one cow. The time you save milking must instead be spent cleaning the machine; it has many parts, and all of it must be scrupulously cleaned after every milking. Every dairy does this.
- If you use a machine you will have mastitis problems. Teat trauma from overmilking or incorrect vacuum due to worn inflations (the rubber

teat cup liners) or vacuum pressure variation in the line predisposes to mastitis. So do lapses in sanitation of the equipment. Over time, you will certainly learn to minimize mastitis, but it is a constant risk and the curse of the commercial dairy. You may have to treat your cow with antibiotics, and then the milk must be thrown away for several days. (Inadvertent failure to discard milk from treated cows is the source of antibiotics in the commercial milk supply.) Additional suggestions on prevention and treatment of mastitis will be found in chapter 10, "Your Organic Cow," and chapter 16, "Diseases and Disorders."

Learning how long to leave the unit on the teats is critically important. Leave it too long and the teats will be damaged, predisposing to mastitis. Leave too much milk behind, mastitis again, and impaired production. I consider seven minutes to be the maximum time to leave the machine on my cow. If she is not milked out in that time, the vacuum is too weak, the pulsator timing is off, or she isn't letting down. With a modern pulsator the timing is built in. The pulsator makes a steady tick-tock. It should make approximately sixty "tocks" per minute. Vacuum pressure is best at about twelve pounds Hg (maximum fifteen) on your vacuum pump gauge. I advise installing an inline pressure gauge. But even without one if the pulsator is behaving properly you can assume your pressure is within a safe range.

Many people find it possible to do other tasks while the machine is on the cow, and in some circumstances this is safe to do. I prefer to stay seated next to my cow. I want to monitor each quarter, often massaging it to get the milk down. I become familiar with the way each quarter feels and know when it is empty. I then remove the teat cup by sticking my finger in next to the teat to break the vacuum before pulling off the cup. (In-line valves are available to perform the task of cutting off the vacuum to individual cups.) At this point I must support the claw with my other hand or the whole thing will fall off due to air entering the line. There are plugs readily available to stopper each cup as it comes off. By catering in this way to each quarter I strictly avoid leaving the machine sucking on an empty quarter, as this can injure the sphincter that closes off the teat and will stress the entire teat. An injured sphincter is slow to close and offers a royal road for access of bacteria. It is advisable to keep the cow standing for at least ten minutes after milking if there is any likelihood she will lie down in manure. This gives the sphincters time to close.

Another source of infection is backdraft of milk up the teat, which can result from fluctuating vacuum pressure. After I remove the machine I hand milk (strip) the last bit from each quarter to wash down any bacteria that may have found their way up the teat. These details of management go a long way toward avoiding mastitis.

There is a milking machine marketed by CoPulsation with a differently designed inflation and pulsator that studies have shown result in less teat trauma and consequently a significantly reduced incidence of mastitis. If you don't want to invest in a whole new system, key components can be purchased separately and will merge with older systems (see the list of sources at the end of this book). Milking machines are by no means the only cause of mastitis. Engorged udders at calving time are another invitation to this highly inconvenient and often costly condition. However, if you have often dealt with mastitis, you may be ready to invest in a design advance over equipment that has changed little in a hundred years.

For machine milking, prepare the udder in the same way as for hand milking. Wash with warm water containing vinegar, rinse with plain water, and dry thoroughly. Unless milk is already streaming out, discard the first couple of squirts. It is important that the cow's teats be full and puffy to ensure that she is letting down before attaching the machine. No force on earth will drag milk out of an udder that is not letting down.

Strain this milk just as you would that milked by hand. The teat cups are little vacuum cleaners and may suck up bits of hay or cow hair.

Suggestion: A full milking machine is both heavy and awkward. A physical therapist made a big difference in my life by instructing me in the best way to lift. Try this: knees bent, feet apart, spine perfectly straight, rear end sticking straight out (not tucked under), and lift straight up. Following this procedure I can lift the full machine the necessary fourteen inches into my cart.

There are milking-machine cleaning units available to speed the washing up. They do a great job but are expensive. Most are designed for the upright DeLaval-style machines. If you have hot water in the barn, all you need to do after straining the milk is reconnect the vacuum hose, set the claw in a bucket of warm water, turn on the pump, and watch the claw suck up the water. Don't put this milky water into your plumbing system or you will eventually build up milkstone in your pipes. This will cause a blockage immune to drain cleaners and baffling to your average plumber. Donate milky rinsings to your pig or your flower bed, where they will be properly appreciated. Repeat with hot water containing a small amount of low-suds detergent and a final hot rinse.

I don't have hot water in the barn. I do the straining, jar filling, and washing up in the kitchen, where I use a wet/dry vacuum (like a Shop-Vac) that provides sufficient pressure to do a good job. Note that water does not enter the vacuum cleaner receptacle. The machine merely provides power. The water follows the same route taken by milk and cleans all the surfaces, pipes, and bucket. I just set the claw unit in a bucket of warm water and connect the vacuum port to the cleaner's sucking hose exactly as I would attach the air line to the vacuum system in the barn. Son Max devised a vacuum hose conversion fitting using the cut-off neck and shoulder of a plastic pop bottle. It holds together easily with the vacuum running. After washing and rinsing, I disconnect one end of the milk hose so that it can drain and dry thoroughly. At least once a week the pulsator must come off so that its port can receive additional cleaning.

A surge bucket milking machine is not adapted to the above method of cleaning unless you splice in hose extenders. Many people use extenders so that the bucket can sit off to the side rather than hang on the surcingle. With the Surge I find it easiest just to pull off the teat cups and lid (removing the pulsator, which must not be immersed) and put them in the dishwasher. The bucket is easily washed with a brush.

Straining the Milk . . . without Delay

Have your milk strainer set up before leaving for the barn, and always strain the milk. There are several reasons for this, foremost among which is that if you don't strain it, somebody in the family who is the fussiest is sure to be the one who finds a cow hair. Such things are not health risks but can quickly erode support for your cow. And there are other important reasons for straining. Any foreign substance will impair the keeping quality of the milk and lead to off flavors. The speed with which milk flows through the strainer and the appearance of the strainer afterward are among your most useful clues to detecting a developing mastitis problem. If the milk strains slowly or if lumps are found on the strainer, this means trouble. However, if the milk has been allowed to cool before straining, this also will make it strain more slowly. It should be strained while still warm, no waiting. If warm milk is very slow to strain *and* you find lumps on the filter, it is best to boil it or feed it to animals and to institute mastitis treatment without delay. A few small, rather tough lumps will occasionally show up on the filter, but by themselves they don't mean much.

If you find a few dried particles of debris on the filter after straining, it doesn't mean the milk is unusable, but it does mean you need to review your arrangements in the barn.

Milk Straining Equipment

There are stainless-steel vessels made for straining milk. They are usually in two parts, a large bowl and a piece to clamp down on the filter disk. The filters themselves are disposable nonwoven material, usually cotton based. They come in boxes of one hundred and can be purchased at feed stores or online (see the list of sources at the end of this book).

A stainless-steel strainer is expensive. You may be able to get a used one by advertising on a local bulletin board or on an Internet site such as eBay or Craigslist. Don't compromise with a flaky old tinned one. It will be impossible to sterilize and may impart a metallic taste to the milk. In a pinch you can strain the milk through a *wet* white linen napkin fixed over a bucket with clothespins. Rinse the napkin in cold water immediately after inspecting it for debris, then wash it in suds and rinse thoroughly. Then either boil it for five minutes or dry it in the sun (a couple of hours in outdoor sunlight will leave anything properly sterilized).

Actually, there is nothing inherently superior about the stainless-steel strainer with disposable filter apart from convenience, and it is an expense. But the linen napkin must be scrupulously sterilized as described, or people will soon be complaining of off flavors in the milk and poor keeping qualities.

If you have only a small amount of milk and believe it to be pretty clean, you can strain it through a very fine nylon or stainless-steel mesh strainer. If you chill the milk rapidly and use it up fast, you may not even notice the difference from milk strained through proper milking equipment. But I promise you, under a microscope you would see plenty of hiding places for bacteria and milkstone, a calcareous residue, in the mesh.

For sustained best flavor and keeping qualities of milk, rapid chilling is essential. In the old days this was usually accomplished by setting the container of milk in cold, running water. A farm with a spring near the house was fortunate indeed. The family could then build a milk house over the spring and divert the water into a stone or slate sink, where it ran continuously, efficiently chilling the milk and creating cool surroundings. If milk was to be set for cream to rise, the bowl would stand on a stone or slate shelf. I am fortunate enough to have in my farm kitchen a granite sink served by a continuously flowing spring. Nonetheless, I usually pour my milk into one-gallon glass jars with lids and keep it in my refrigerator. Keep the refrigerator at 40°F or lower.

Washing Up

The way you wash up is more important than what soap or chemicals you use. But in general, a low-suds detergent is preferable as it rinses more easily and completely. There are dairy chemicals on the market with which you can do a final rinse of all utensils. You leave this rinse on until the next milking, and then rinse with plain water before using. If you are using a machine, it may very well be worth doing this, as there are so many crevices in which milk residues can hide. Many of these sanitizing agents contain iodine, and on some commercial dairies enough of this finds its way into milk to make milk a significant iodine source. Other agents are chlorine based and if not completely rinsed away will destroy vitamins E and C and any other antioxidants with which they come in contact. Nonetheless, they are unquestionably effective germicides and have their place.

I use low-suds soap and water and the following procedure:

1. Rinse first with lukewarm water. This is critically important because hot water solidifies milk protein and makes it stick like Elmer's glue, in which milk casein is indeed an ingredient.
2. Follow up with a thorough scrubbing, using hot water and a nonsudsing detergent in a dishpan that is not used for anything else. I use a big fluffy dairy brush with white bristles. Dairy brushes are pricey but last for many years, and this brush is never used for any nondairy purpose. I use this brush to scrub only surfaces that come in contact with milk. It never touches the outside of buckets that have been to the barn. I pour soapy water into the bucket that has been to the barn so that its unclean outside is not immersed in the basin used for milk surfaces.
3. Rinse thoroughly with scalding water.
4. Set out to air-dry in the sun or somewhere with good airflow. Indoors or out, I cover everything with clean netting.

This procedure maintains something approaching sterile technique with everything that comes in contact with milk, and I know it works. I know it works because milk doesn't lie; mine will keep more than two weeks in the refrigerator without going sour or even significant loss of flavor. I've known it to keep three weeks, which is as long as commercial pasteurized milk is expected to keep in most states unless ultrapasteurized.

I usually wash milk jars in the dishwasher, after rinsing them with lukewarm water. I routinely rewash jars that come back to me from customers. Whenever milk doesn't keep, it will be the milk that gets the blame, not the jar. Not fair, but that's how it is.

How Much Water Does All This Take?

It is possible to get along with only two or three gallons of water if there is a shortage or if it must be hauled. Save the cloudy water from the initial lukewarm rinse for another purpose. It is perfect for pigs. It will cause your houseplants to thrive amazingly. With your fluffy dairy brush you can wash the milk-contact surfaces with only about two quarts of hot water with low-suds detergent. Boil any of the water that remains after scrubbing and pour it over the milk-contact surfaces. All of this water can be saved for other wash-up tasks.

A milking machine demands more water for thorough cleaning.

Must I Really Do All This Washing Up?

After all, in many parts of the world it would be simply impossible. Is somebody going to get sick if I skip it? No, they will not. But there are two parts to this answer.

If the milk is used immediately, off flavors and keeping qualities are not issues. In places without running water or refrigeration, milk not to be consumed immediately is preserved by various types of controlled lactic fermentation that result in a tasty product. If freshly drawn milk is put into unwashed vessels dedicated to milk and milk alone, a favored colony of lactobacillus can be maintained. Prior to modern refrigerators, even if milk was in clean containers, nobody expected fresh milk to keep more than one day without going sour, even if it sat next to a block of ice, at least in warm weather. The milkman had to come every morning.

Now, the second part of the answer. If there are milk customers they will expect their milk to keep at least a week. If the milk is to be used to make yogurt or cheese, you will want to ensure a very low bacterial count, because inoculants may not compete successfully with a huge population of wild bacteria, and the quality of the resulting cheese or yogurt will be inconsistent. So strict sanitation must be observed: clean milking, careful washing, and rapid chilling. Milk stored in covered containers under refrigeration favors a type of bacteria that results in bitter flavors. This will be the ultimate fate of even the best-cared-for milk Indifferently cared-for milk may become unpalatable after as little as twenty-four hours.

"Unpalatable" does not mean it will make you sick. Hazards from commercial food supplies have been much publicized and are indeed serious. The pathogens in commercial food supplies are easily avoided in home-produced foods. The worst are not present at all. The old-fashioned common bacteria of everyday life will always be with us but are not harmful. They just make the milk go sour. Careful milking and milk handling practices keep bacterial start-up numbers low; refrigeration slows or prevents their proliferation. Inoculation of milk with favored bacteria such as yogurt starter suppresses the growth of bacteria you don't wish to encourage. Pathogens do not survive in the acid environment of fermented milk products. We take pains with home-produced milk so that our dairy products will be dependably delicious. Happily, with such precautions they are also dependably safe.

Rising Cream

To obtain the most cream without use of a cream separator, pour the warm milk immediately into a large, flattish bowl and cool it. A cream pan is traditionally shaped like a giant pie dish: flat bottom and sloping sides. The less handling and the less time that elapses between milking and the time the milk is set to rise, the greater will be the percentage of cream you get. For the thickest cream, go for this wide, flat cream pan. The skim milk that is a by-product can be used for animal feeding or scalded for bread making.

If you're planning to serve whole milk and prefer not to remove all the cream, pour the milk into one-gallon food service jars. This milk will produce more cream than even a dedicated whole-milk user such as myself wants to drink. Just pour or ladle off some of the cream before serving. This will give you a supply of cream for butter, desserts, and coffee. Cream poured off a jar will usually not be as thick as cream skimmed from a pan.

I often find it convenient to pour the morning milk into jars and set the evening milk to rise in a pan. In either case, for cooling I usually refrigerate the milk. I find that with ruthless elimination of leftovers, I can get a big bowl or pan into my refrigerator. I cover it with a tray to avoid condensation from the shelf above. In cool seasons the pan may not require refrigeration.

Temperatures of around 50°F will permit the cream to ripen slightly while rising. In the opinion of most, ripening improves the flavor of butter, and this "ripened" cream will break into butter more quickly.

If There Doesn't Seem to Be as Much Cream as You Expected

As a rough guide, in a glass food service jar or any straight-sided jar, you should be able to see at least two inches of cream by the time the milk has stood twelve hours. I am disappointed if I don't see a three-inch layer. By the following day the cream yield of certain cows will be even greater. Spring grass always boosts the amount of cream. There are genetic factors affecting the percentage of cream. Few Holsteins are big cream producers, for example, and there are feed factors besides grass that influence the amount (see chapter 9, "Feeding Your Cow"). But there are some simple management details that cause striking variations in the amount of cream. Here are a few:

- Incomplete milking: The hind milk – that which is milked out last – has a much higher cream content than the earlier milk. This seems to be the case for all mammals. If whoever is milking the cow is, let us say, racing to catch a schoolbus, that extra time spent getting the last of the milk may be skipped. The result, besides declining production, will be less cream.
- Feeding the calf . . . or the cats: Even while you are milking, the cream is rising fast in the pail. If before straining you pour off a share for the calf, or even a couple of pints for the cats, you are pouring out the creamiest milk. If you are straining directly into jars you will note that the jar you fill first ends up having the most cream.
- Extra agitation: Lots of pouring back and forth of milk and use of some types of milking machines causes partial homogenization by disrupting fat globules. Cream then rises more slowly and doesn't get as thick. You still have the cream, but it's stuck in the skim milk.
- Allowing the calf to suckle: If your management scheme involves allowing the calf access to the cow after partial milking, leaving some in the udder for the calf, this will dramatically reduce the amount of cream you have in your share. This is not only because the calf is getting the hind milk. If the cow has reason to believe that she will be suckling the calf later on, whether on a regular or erratic basis, she will resist letting down

her milk. Whenever a cow isn't letting down, cream is preferentially held back by the udder.

- Stress factors: Slow, reluctant letdown or cessation of letdown during milking can occur due to any of the disturbances mentioned earlier, such as the presence of a dog. Imperfect letdown will diminish the amount of cream in the milk.

Cream Separators

New cream separators are available. Good old ones can often be found. There are hand-cranked and electric models. The parts that come in contact with the milk are usually stainless steel and so are resistant to rust. Like a milking machine, cream separators have a great many parts that need to be scrupulously washed. But they work extremely well, so if you want to make lots of butter you may find a separator to be worthwhile. A separator centrifuges the milk in sheets as it passes over its many conical disks. The cream comes out one spout, the skim milk out another. There is a screw adjustment that regulates the thickness of the cream.Your fresh warm milk is strained directly into the hopper after milking so you gain the advantage of being able to deal with the milk immediately and completely – no refrigerator full of jars or pans. You do, however, forfeit the delicate flavor of milk and cream that hasn't been bashed around.

The most perfect milk can only be had from your own cow, and this for me is a major reason for having a cow at all.

CHAPTER 4

Making Butter, Yogurt, and Cheese

In the days of Good Queen Bess the cook often used to send a maid out to the dairy with half a glass of wine to be topped up with milk directly from the cow. This posset was believed to restore health and friskiness to the weak and weary. Most people nowadays like milk well chilled. I let mine stand in a cold place overnight and usually skim off part of the cream before serving it. Fresh milk whether warm or cold should have a light, sweet flavor. Well-flavored, strictly fresh milk is essential to the manufacture of the best milk products.

Of the vast treasury of milk products representing nearly every civilization in the world, I will discuss only a few. Best known in the American tradition are cream, butter, buttermilk, yogurt, cottage cheese, cream cheese, and hard cheese. If you are not already acquainted with ghee, do try it.

Cream

Cream obtained by skimming milk that has stood for twenty-four hours is an incomparable product. Over apple pie it is an unforgettable experience. For a long time cream held out against the food technologists and was a better buy than other commercial dairy products, but no longer. Real cream is now almost unobtainable. The cream now sold as all-purpose, heavy, or whipping cream is artificially thickened by the addition of alginate, a mucousy although (so far as I know) harmless substance. The resulting "cream" has an uninviting cooked taste reminiscent of canned milk. If you want real cream (and you will, once you've tasted it), you must have access to a cow. Whipped unpasteurized cream is luscious. You'll understand right away why the word "cream" has entered the language as a form of praise.

After your milk has stood for twenty-four hours the very thickest cream will be at the top and underneath will be the lighter cream, which may not whip. To skim the cream by hand, slip a perforated skimmer or a ladle under the heavy layer of cream. Too high a proportion of light cream in your churn will interfere with butter making, so you may wish to create a "coffee cream" category.

If your cream has come through a separator, it will be of uniform consistency. It will still be warm and may not seem very heavy, but it will thicken significantly after a day of chilling. Separated cream can be made into butter as soon as it is chilled if this best suits your management.

Making Butter

Sweet cream can be made into butter, but most people find the flavor to be more interesting if slightly "ripened" cream is used. Ripened or slightly sour cream also takes less time to turn into butter, which is a consideration if you are churning by hand. Ripened cream of consistent flavor can best be obtained by inoculating the cream either with mesophilic starter or with sour cream starter. Starters (inoculants) can be obtained from a cheese-making supply house (see the list of sources at the end of this book). Commercial buttermilk can be used as starter if it contains active cultures – many do not. If you are in the habit of making clabber and it has a good flavor, this works fine too.

To make butter, you'll need plenty of cream, which usually means that you'll collect the cream from at least a few milkings. Save it up in the refrigerator until you are ready to make butter. (Some electric churns come with a large glass jar that holds the cream. Simply pour your cream into this jar until it is about half full; there should be a fill mark on the churn.) If you're using starter, stir it into the final addition of cream. With or without starter, leave the whole lot to get acquainted for two to four hours before churning. Omitting this step will make the cream difficult to churn and leave some of the cream behind in the buttermilk.

Cream usually takes about twenty minutes to churn, or longer if it's too cold. It churns better and faster at near 64°F. To quickly warm cream for churning, a wet, hot towel can be draped over the churn. If the cream is too cold, it will take virtually forever to churn and will probably begin by first becoming whipped cream.

An electric mixer can be used for small quantities. Mine is efficient for this purpose, using the batter beater rather than a wire whisk. If you do get whipped cream, just keep beating, and it will finally become butter.

For several years now my collection of churns has stood unused. I have discovered that my food processor fitted with the plastic dough blade makes butter more efficiently than anything else. It rarely takes more than three minutes, and usually about 90 seconds. I simply churn the butter in the food processor in batches, pour each batch through a cheesecloth-lined colander, and drain and wash the butter in this.

Use of the metal blade in a food processor or blender has proved problematic. Equipment varies, and yours may make good butter, but I and many readers have found that a metal blade results in butter that rapidly develops the off flavor characteristic of rancidity. I suspect that the metal blade damages the fat globules, permitting rapid oxidation.

If you prefer to use very basic methods, you can just shake the cream in a jar, take it for a ride in a car, or drive it around on your tractor to keep it jiggling. The lid must be opened at least twice to allow pent-up gas to escape. A proper churn always has air vents.

If your cream is slightly sour, on the way to becoming butter it will often pass through a thickened stage about the consistency of commercial whipped topping. If you stop the churn at this stage, you can save some of this product, which is perfect sour cream of the sort favored for cheesecakes and baked potatoes. With hand churning this stage does not occur.

I think hand churning is dreary business. If you must do it, be particular about cream age, blend, and temperature, or it can take forever. No matter how determined you are, occasionally you will find the cream is inexplicably resistant to becoming butter. Sometimes it seems to get halfway and stop. Often a dash of cold water added at this point will cause the cream to "break." In former times women were sometimes accused of witchcraft on the basis of a housewife's unsuccessful butter making. There is an unpredictable, mysterious element in churning that makes it easy to understand how people might reach this unfortunate conclusion. Witchcraft aside, here are a few factors that can cause trouble and are worth knowing about even though you can't do much about them: In late winter cream is slower to churn than in June, when you may get butter in three minutes. Cream from a cow in late lactation is often slow to churn. And there are certain cows (rare) whose cream is always resistant.

It is possible to overchurn when using an electric churn. If the cream is too warm or if you get bored and wander away, the butter may come and then be beaten back into the buttermilk. The resulting product looks something like butter but is softer and whiter, almost like mayonnaise. No amount of beating will restore real butter. You can rescue it pretty well by adding very cold water to harden the mass and kneading it for a long time, first under water, then drained. It will resemble the butter substitutes some people make by adding oil to butter or that fake spread sold in a tub. It won't keep very well. But don't throw it away. Maybe somebody will want it as "low-calorie butter." Or you can melt it down for ghee (see page 63).

Washing and Pressing

After about ten to twenty minutes of churning (except with a food processor, which is much faster), little yellow flecks should appear. You will note that the sound of the churning changes as the butter breaks. When these lumps are about the size of a pea, stop churning and pour the lot through a strainer, saving the buttermilk. I use a piece of (wet!) butter muslin (finely woven cheesecloth) over a large bowl or colander. Butter muslin is hard to find. Try a net curtain or order fine mesh from a cheese making supply catalog (see the list of sources at the end of this book). Holding the cheesecloth like a hammock, rinse the butter in at least three changes of water until the water remains fairly clear. The water must be colder than 60°F, preferably in the range of 45°F to 50°F, or the butter will become too soft, stick to the equipment, and be unworkable.

After the butter is washed, it must be pressed to remove all water. Before pressing, add one teaspoon of salt for each pound of butter, unless you prefer it unsalted. Press the butter thoroughly by folding and pressing, never smearing, which will develop a greasy consistency rather than the waxy consistency characteristic of good butter. Keep going until water is no longer being squeezed out. All the buttermilk and water must be out if the butter is to keep well. If working with more than a pound, I can't press the butter with a wooden paddle. In that case I chill my hands under cold water and do it with my fists. With smaller amounts or when the butter is quite warm, a butter paddle (presoaked in cold water) works nicely and saves having to answer the phone with buttery fingers. Butter can be pressed on a wooden board or in a wooden bowl.

Once the butter is pressed, pat it into any shape you like and wrap it in butter paper, if available. I dislike using anything but butter paper, but it can be very hard to find.

I have a form that holds one pound. I make pound packets and sometimes sell part of it. The market for homemade butter is strong at all times. It is also a much appreciated gift. As a trade item it may be hard to get a fair return. As you can see, butter is a lot of work, and you may not be able to sell it for as much as it's really worth. But when you put homemade butter on the family table, you will wonder how you ever settled for commercial butter, that pallid, disinfected product. From six quarts of cream, I usually get five pounds of lovely fresh butter.

Vitamin E

I add about 200 IU vitamin E per pound of butter. It can be added to the churn or to the butter when pressing. I have done this for years because I have found that it effectively prevents rancidity. Homemade butter "goes off" much sooner than commercial butter washed using industrial methods. Butter fortified with vitamin E and stored in the freezer will still be edible a year later. Otherwise three months is about the limit. If you are selling butter you will especially welcome the better keeping quality. Just puncture a vitamin E capsule and add it to the cream before churning or work it into the butter while pressing it.

Ghee or Clarified Butter

Ghee is butter that has been melted and clarified by draining the clear oil away from the remaining white sediment of milk solids. If you live in a very warm climate you will know why ghee is preferred in India (its country of origin) instead of butter. Warm butter seems determined to become ghee. Ghee has merit in northern climates too, though, for its convenience; it is less trouble to make than butter. Just churn your cream, strain off the buttermilk, and rinse the butter in cold water. Here friend Lee Anne B. describes her method:

> *Basically, you put butter into a stockpot and heat it up. I use medium to medium-high heat. Don't cover the pot. The butter will melt, then boil. When it's first boiling you have to be aware and watch for it to avoid boiling over. Adjust your heat to ride the fine line between "fast" and "out of control." At one stage, it will have a thick layer of foam on the top. Stir it every once in a while. Eventually, the moisture will boil out and the milk solids will find each other and solidify into curds. At this point, start paying closer attention and stirring more. The liquid will go from cloudy looking to a perfectly clear golden color, and you will be able to see the bottom of the pan without obstruction. The curds will sink and want*

to stay at the bottom of the pan. At this point you are almost done, and you need to keep stirring. When the milk solids are a golden brown, take the pot off the heat, and pour the liquid through a filter (a milk filter works well) into jars. As it cools, it will solidify into a "soft solid" – it's about the consistency of coconut oil, but smoother.

During the "foamy" stage, some people say to skim the foam. Some people say to stir it back in. What I do is turn the heat up just a little bit and let the boiling action work it back in. Skimming, to me, wastes a lot of the oil and makes a bigger mess. It also takes more time and attention, and I'm all about the easier thing.

I use wide mouthed pint jars – they are the perfect size and shape to get the ghee out with a spoon or knife, and the right depth to be able to get to the bottom without having to get your hand in a jar. I keep the extras in my freezer. The jar I'm using stays right by my stove.

Ghee has excellent cooking qualities. Unlike butter, it does not scorch readily. It lacks much of butter's flavor but has a nice flavor of its own. Use ghee as you would any fat or oil in frying or baking. Any cream, young or old, can be used in making ghee. It is a satisfactory way to preserve butterfat for a later time. It can be frozen but will keep nicely on the pantry shelf or can be sealed in hot sterile canning jars. The jars need no further processing. It will keep this way for a very long time, and even longer if fortified with vitamin E.

Lee Anne B. again: "One of my favorite things to do with ghee? Slice squash, dice onions, slice mushrooms, mince some peppers, dice some tomatoes, chop up some collards, kale, chard, or other greens, and cook the whole mess in ghee, and season with just sea salt. Be liberal with the ghee, and when the veggies are done, there should be a delicious, golden, buttery sauce. Oh *wow*. So, so good. My dearest, who doesn't like veggies, eats every bite, gets seconds, and drinks the broth out of the bowl. I smile as I count vitamins and omega-3s. You could of course, use bigger pieces or slices, but I have to hide some things. Two of my 'boys' hate tomatoes and greens. I laugh every time they gripe at me for not ever making enough of this dish."

Buttermilk

When cream is agitated to make butter, it splits into two parts. The solid fraction is the butter and contains nearly all the butterfat. The liquid part is the buttermilk, with properties similar to those of skim milk but with its own special flavor due to the presence of diacetyl, a ketone. It contains milk proteins and sugars and is valued for its superior baking properties. Buttermilk from butter making is not the same as the product sold in stores. Commercial buttermilk is a product concocted from powdered skim milk and thickening agents. I suppose it's all right for cooking, but you would not want to drink it. Your genuine product will usually be delicious. Buttermilk from sweet cream has less flavor, while that made from sour cream will be tangy.

The appearance of your buttermilk may sometimes be odd; if it separates into curds and whey, just shake it up again. If upon standing a thick layer of cream rises on the buttermilk, you will want to review your butter-making procedures. It usually means that when you were saving up cream for the butter, you didn't stir in the last addition and allow it to rest and become unified. It can also mean that the churn was too full. If the cream cannot circulate freely, the bottom half of the cream will turn to butter before the cream on top. (If you notice this happening, open the churn or food processor and mix it down with a spoon.)

If the flavor of your buttermilk is disappointing, there is a trick for improving it. It sounds counterintuitive, but it works. Save half a cup of the poorly flavored buttermilk and add it to the cream the next time you churn. Repeat this each time you churn and the flavor will progressively improve.

If you have tasty clabber going, try adding some to your cream before ripening. This should ensure good flavor in the buttermilk. It is also possible to order buttermilk starter (similar to yogurt starter) from cheese-making suppliers.

Making Yogurt

Milk from a high-milk-solids breed such as the Jersey makes the best yogurt. Use whole milk and you'll get a little "frosting" of yogurty cream on top. I often make mine in eight-ounce ceramic pots, glass jelly jars, or coffee mugs. Sometimes I just fill a large pan with milk.

I have uneven success with yogurt unless I first scald the milk to destroy the competing bacteria, but this does forfeit some of the essential value of raw milk. The important point is that the milk must be between 90°F and 115°F to make yogurt. So you can scald your milk, then let it cool to this temperature, or if you don't want to destroy the probiotics in raw milk, you can simply warm your milk to this temperature.

For yogurt starter, for every two quarts of milk, use about one-quarter cup (no more; the organisms don't like to be crowded) of your last batch of yogurt or of a first-class brand of fresh plain yogurt, or use a commercial starter and follow the packet directions. When your milk is warm, add the starter, and stir it in gently but thoroughly with a very clean spoon. Prewarm the vessels in which you plan to set your yogurt, then pour or ladle it in and cover them. Find a spot in your home that will keep the yogurt cozy, without allowing it to fall below 90°F or heat it above 115°F. It is worth a bit of trouble to find the ideal spot for incubating your yogurt. Once you hit on the successful combination of container and heat source, you can make yogurt quickly and easily, without having to give it much thought. The incubation should take about four hours. Try not to jiggle the yogurt while it is setting or it will be runny.

If your yogurt is not as thick as you would like, there are several things you can do about it. To begin, note that raw milk yogurt tends to be less firm. To firm it up a bit, you can let it sit longer, or you can obtain fresh starter. Using a larger amount of starter is unlikely to be helpful. The yogurt organisms prefer not to be crowded. Do be sure to mix thoroughly. I usually make yogurt in one big pot and ladle out the whey as it accumulates. This has the effect of thickening the yogurt. You can make Greek-style yogurt by draining the whole batch of yogurt through a wet linen tea towel to remove some of the whey.

Crème Fraîche

Crème fraîche is pretty much like sour cream. I would call it a distinction without difference. Either way, it's wicked good, as we say in Maine. The most reliable way that I have found to make it is to inoculate heavy cream with mesophilic culture, a cheese-making culture available from cheese-making supply houses (see the list of resources at the back of this book). Follow the directions for the mesophilic starter and you will end up with a couple of quarts of a yogurtlike clabber. I pour this into ice cube trays and freeze it so that it is conveniently on hand. One cube in a quart of cream is all you need to make crème fraîche. Stir the cream thoroughly after the cube has melted. Set the jar somewhere that maintains warm room temperature, about 70°F, as for bread. It usually takes less than eight hours to thicken.

You can start your next batch with a tablespoon of crème fraîche from your first batch. This will work well for several batches, but then you'll need to use a new frozen cube. If you already have on hand some well-flavored buttermilk or clabber, this also makes good starter.

Cheese Making

There are few civilizations past or present in which dairy animals are unknown. Everywhere that dairying occurs, cheese making also has evolved. The natural tendency of all dairy species — sheep, goats, cows, reindeer — is to bear their young in the spring. There is then plenty of milk for all purposes. Later during the winter there may be no milk available at all. This is not because milk cannot be produced during the winter but because hay making is hard work and there was rarely enough to keep a cow or other animal sufficiently well fed to support much winter production. So the best time for cheese making is during the spring, when the young animals appear. Cultured dairy products are essential for the preservation of milk. Yogurt and allied products serve to preserve milk for a few days. Hard cheese, with its wonderful keeping qualities, has been the natural answer to milk preservation nearly everywhere.

During cheese making, milk sugar (lactose) is converted by the enzyme lactase into lactic acid, much of which remains in the whey. Calcium, being soluble in acid, also tends to be lost with the acidic whey. Fat is present according to the amount of cream in the milk. Most of the fat and protein in the milk remains in the cheese.

A virtue of cheese and other cultured milk products is that disease organisms, whether from ill animals or from contamination of the milk during handling, are effectively destroyed by lactic fermentation. Virtually all harmful bacteria share the characteristic of being unable to survive in an acidic environment. While this fact should never make you careless with your own milk supplies, it nevertheless offers an explanation for how it has been possible for many people, without knowledge of sanitation, refrigeration, or much of an understanding of disease, safely to use their dairy products. Aging the cheese further destroys bacteria and improves the digestibility of the milk protein.

In order to make cheese, milk must clot. Milk clots (forms into a solid) owing to the precipitation of casein (milk protein). Humans have learned to have some control over this transformation using either the enzyme rennin, which provokes a rapid response, or acid, with a slower one. Rennin is present in the stomachs of young mammals to help them digest their milk. It is extracted from calf stomach tissue for use in cheese making. Plant derivatives may be substituted. Acid curd (as in yogurt or clabbered milk) is a result of the breakdown of lactose (milk sugar) due to the action of beneficial bacteria. You can hasten and control this effect by adding a small quantity of known bacteria, called a "starter" or "culture," as is always done in yogurt making. Or you may depend on wild bacteria that quickly enter the milk after it leaves the cow, in which case the flavor will be unpredictable.

Once the milk clots, the protein begins to shrink and toughen. The liquid portion, whey, begins to separate out. What remains behind is cheese. How hard it becomes, the flavor it develops, and its keeping qualities depend on how it is handled from this point forward.

The whey from any cheese making should not be discarded. If you have pigs you already know how they love it. Diluted, it makes impressive fertilizer. And of course there are culinary uses for whey; it is excellent for bread making.

Soft Cheeses

To get your cheese making started, here are instructions for a few kinds of soft cheese. The soft, spreadable cheese made with yogurt is easiest and is a tasty way to meet the protein requirement of anyone who doesn't like eggs.

Cottage Cheese

There are a number of ways to make cottage cheese. Here is my favorite.

Put skim milk (I favor raw milk) in a large bowl or gallon jar. Amounts are not important, but use around a gallon. Add about one-half cup of starter culture, stirring well. For the most predictable results, use mesophilic starter. However, cottage cheese is very forgiving. You can use anything you have that contains an active culture. Buttermilk will work, and I like to use well-flavored clabber (milk that has thickened naturally). If you have no starter, just rely on wild bacteria.

Cover the milk to exclude light, which destroys the riboflavin in milk and also alters the flavor. Make sure the cover is loose rather than tight-fitting, though, or the milk may develop off flavors. Set the bowl or jar where it can maintain warm room temperature, 65° to 70°F. Your milk will form into a solid curd called clabber within twelve hours (if starter was added) or at least by the second day (without added starter).

After the milk forms a curd, gently break it up into smaller curds. While very gently stirring, heat the curds to 90°F to 110°F. The lower temperature is preferred by those wishing to preserve all the special properties of raw milk; the higher temperature will give you the firmer crumbles characteristic of dry cottage cheese.

Next, ladle the curds into a colander lined with wet cheesecloth and allow them to drain. The curds of the low-temperature cottage cheese are so soft that it is best drained through a double layer of cheesecloth or high-quality paper towels. When well drained, turn the cheese into a bowl, salt to taste, and add some cream or melted butter if desired. A scoop of this along with beans and corn bread will give you a meal of outstanding protein quality at very low cost.

The temperature of your cottage cheese after you have heated the curd will give you a reference point for your future efforts. If your cheese turns out too dry or hard for your taste, heat to a lower temperature next time. If it is too soft, raise the temperature a few degrees.

Neufchâtel

Whole-milk yogurt can also be drained in this manner to create Neufchâtel, a lower-fat cream cheese with excellent flavor. I have used it with great success in cheesecake and pashka (Russian Easter cake). It is also an excellent spread. It is a more honest product than commercial cream cheese, which contains considerable vegetable gum to create its firm texture. One of my daughters declares that a scoop of Neufchâtel in a bowl with cream is her favorite snack and will sustain her through hours of work.

Quark

The preparation of quark varies. At its most basic, drain clabber overnight until thick enough to handle well in cheesecake or as a crêpe filling. For a firmer product, heat clabbered or otherwise cultured milk very gradually heat to about 100°F. Drain in a linen tea towel until it reaches the desired thickness.

Cream Cheese, the Real Thing

Combine one quart of heavy cream with two ounces of mesophilic culture (or buttermilk). Dilute one drop of liquid rennet in two tablespoons of water, and add that to the cream. Set aside to culture at room temperature for twenty-four hours.

Ladle half of the curd into a colander lined with wet butter muslin. Sprinkle one teaspoon of coarse salt over the curd in the colander, ladle in the other half, and sprinkle the top with an additional teaspoon of salt.

Make the butter muslin into a bag by tying up the corners, and hang it to drain for twelve hours.

After draining, remove the cheese from the butter muslin and place it onto a clean piece of wet butter muslin. Press at five pounds or so for six hours, then chill. It keeps for about a week in the fridge.

You can skip the pressing if you shift or turn the cream cheese around in its bag a couple of times for better draining. After draining, put it in a cloth-lined strainer and store it in the refrigerator with a plate on it, topped by a slight weight such as a jam jar.

The yield from one quart of cream is about fourteen ounces.

Queso Blanco

Queso blanco is a versatile and tasty cheese that is very quickly made. In texture it is halfway between hard and soft. Don't use an aluminum pan, or the milk will discolor and taste funny, not to mention potential toxicity. You can use whole or skim milk, but whole milk makes a tastier cheese.

Heat 1 gallon of milk to 185°F, stirring gently and constantly with a wooden spoon.

Stir in apple cider vinegar a little at a time. The milk will begin to curdle. Continue to add the vinegar until the whey ceases to look milky and becomes clear. (You'll probably need somewhere between one-quarter and one-half cup of vinegar in total.) Do not boil. Remove from the heat. Pour the curds and whey through a colander lined with wet butter muslin or cheesecloth. Make the butter muslin into a bag by tying up the corners, and hang it up to drain. Within an hour or less the cheese can be sliced and used. By the next day it will very firm.

The vinegar cannot be tasted in either the cheese or the whey.

Queso blanco may be cut into cubes and added to practically any dish. It takes up flavors without melting. Try sautéing it in melted butter flavored with curry powder. Add it to cooked rice or pasta and vegetables for a complete meal. Or sauté slices of queso blanco in butter and serve with maple syrup. I also like to cut queso blanco into small cubes and marinate it in olive oil with herbs and garlic.

I find that Jersey milk requires the larger amount of vinegar. You should get about two cups of queso blanco from a gallon of milk, and the whey should be quite clear. If yield is low and the whey looks milky, heat it up again and add more vinegar for a second precipitation. The whey is useful in many ways. It can be used in making bread or soups or as a beverage.

Hard Cheeses

Making hard cheeses is not really difficult; like most worthwhile enterprises, it is just time consuming. It can be one of the most rewarding activities of your home dairy operation. Cheese making can liberate you from buying an increasingly expensive commodity; cheese can scarcely be considered a cheap meat substitute. Most cheese now sold in the United States except that from organic producers will be made from milk containing bovine growth hormone (BGH). As of this writing most imported cheese will not contain BGH because few other countries permit its use. However, an increasing number of other items may legally be added to cheese, such as imported casein.

Many small dairy operations with two to four cows now make a significant contribution to the household income with their cheeses. Quality homemade cheese sells for a high price and, unlike other dairy products, improves with age.

I do not pasteurize my milk when making cheese. If you prefer to do so, heat it to 145°F and hold it for thirty minutes at this temperature. Then cool the milk quickly by placing the pot in a sink of cold water.

Cheese was made for centuries without the use of thermometers. Accomplished cooks become very accurate about this sort of thing, just as accomplished choristers can remember the opening note given by the director down at the basement practice piano. For the rest of us, technical aids are a necessary crutch. Let your first cheese-making investment be a good dairy thermometer. You'll use it for all your dairy products hereafter, and it will quickly pay for itself. The other item that's difficult to manage without is a stainless-steel container. Unchipped enamel will do, but no other metal. What people did for centuries before they had stainless steel was to use a wooden box. To heat its contents, they would ladle out dippers of whey, heat it over a little fire, and return it to the vat. Standing your container in a larger vessel of hot water or in a sink full of hot water also works. Have a rack between the milk pot and the bottom of the outer container. The water temperature should not exceed the temperature of the milk by more than ten degrees. Stir with a stainless steel or wooden spoon.

Be sure that everything that comes in contact with the milk is very clean, and try to avoid raising dust around your cheese making. Of course you could use wild bacteria for starting your cheese, but a mesophilic (moderate-heat) bacterial culture will produce a characteristic, predictable cheese flavor. You can buy mesophilic cultures from any cheese-making supplier (see the list of sources at the end of this book). For similar reasons, avoid making bread at the same time as cheese; random yeast spores can spoil cheese.

A lot of cheese-making instructions are necessarily devoted to the hardware that's needed, since most kitchens will not have all the necessary materials on hand. For the pressing process, many methods will work. The Lithuanians in my area use a pair of boards hinged together at one end. The cheese is wrapped firmly in a cloth and pressure is applied with a screw arm that locks down the open end and can be tightened; the resulting cheese is wedge shaped. If you prefer, you can contrive your own press using a cylindrical tin. You will need to make a follower, a solid disk that fits inside the cylindrical tin. You put the cheese inside the tin and set weights on top of the follower, and it presses down on the cheese, squeezing out any remaining whey. For weights you can use something as simple as bricks or jars of water, although keeping these balanced can be tricky. Many people have described an alarming midnight crash when the weights shifted and toppled.

When I'm aging hard cheeses, I find it useful to cover them with an old-fashioned fly screen. This is like an overturned sieve. It protects the cheese against marauding cats (they love cheese) and gnats, which will quickly destroy a cheese with their ruinous little maggots.

Different cheeses are the result of variations in the timing of the procedures with the curds and whey, variations in the fat content of the milk, the temperatures at certain stages, and the addition of characteristic bacterial starters. Other variables are the amount of pressure and length of aging. I write down just how I did each cheese and keep my notes with it; for instance, "temperature went to 104°F with this cheese." None of my cheeses has been truly inedible. Some have been a little disappointing, but some have been marvelous. For predictably good results, you should obtain a cheese starter culture and develop the right sort of equipment and storage conditions. Most hard cheeses can be eaten after three to four weeks, but hard cheeses improve with longer aging.

Here are two of my favorite recipes for hard cheeses.

Farmhouse Cheddar

This recipe comes from *Home Cheese Making*, by Ricki Carroll, proprietor of New England Cheesemaking Supply Company, and is used with permission.

2 gallons whole milk
1 packet direct-set mesophilic starter or 4 ounces prepared mesophilic starter
½ teaspoon liquid rennet (or ½ rennet tablet) diluted in ¼ cup cool, unchlorinated water
1 tablespoon cheese salt
Cheese wax

1. Heat the milk to 90°F. (If you are using goat's milk, heat it to 85°F.) Add the starter and stir thoroughly. Cover and allow the milk to ripen for 45 minutes, maintaining its temperature. If keeping the curds warm using a water bath, the water should be 10 degrees warmer than the target temperature.
2. Add the diluted rennet and stir gently with an up-and-down motion for 1 minute. (If you are using farm-fresh cow's milk, top-stir for 1 minute with the flat underside of the ladle no more than ½ inch deep to blend the butterfat that rises to the surface.) Cover and let set at 90°F (85°F for goat's milk) for 45 minutes, or until the curd gives a clean break. Check for a clean break by inserting a thermometer or your finger into the curd at a 45-degree angle. The resulting cut should break cleanly.
3. Cut the curd into ½-inch cubes.
4. Place the pot in a sink full of hot water and slowly heat the curds to 100°F, increasing the temperature by no more than 2 degrees every 5 minutes. This will take about 30 minutes. Stir gently to keep the curds from matting. The curds will shrink noticeably in size as the heating continues and you stir gently. The yellowish whey will grow in quantity as the curds shrink.
5. Cover the container and let the curds set for 5 minutes. Pour the curds into a cheesecloth-lined colander. Tie the corners of the cheesecloth into a knot and hang the bag in a convenient spot to drain for 1 hour. Do not hang in a drafty spot – the curds need to stay relatively warm.
6. Place the drained curds in a bowl and break them up gently with your fingers into walnut-size pieces. Mix in the salt.
7. Firmly pack the curds into a 2-pound mold lined with cheesecloth, then neatly fold the cheesecloth over the top. Apply 10 pounds of pressure for 10 minutes.
8. Remove the cheese from the mold and gently peel away the cheesecloth. Turn over the cheese, re-dress it, and press at 20 pounds of pressure for 10 minutes.
9. Repeat the process but press at 50 pounds of pressure for 12 hours.
10. Remove the cheese from the mold and carefully peel away the cheesecloth. Air-dry the cheese at room temperature on a wooden board (such as a sushi mat) until a nice rind has developed and the surface is quite

dry. This can take 2 to 5 days, depending on the weather. Turn the cheese several times a day so moisture will not collect on the bottom.

11. Wax the cheese (see the discussion below).
12. Age the cheese for at least 1 month.

Mesophilic culture (direct set or powdered culture for making your own starter), rennet, cheese salt, wax, and cheesecloth are available from New England Cheesemaking Supply Company (see the resources at the end of this book). Also available are molds, thermometers, and presses.

Note: To wax a cheese, first melt the wax and pour it into a pie pan. Wax the top and bottom of a molded cheese by setting each side quickly in the wax, and then roll the sides in the wax. There are also instructions for waxing on Ricki Carroll's website (www.cheesemaking.com).

Semihard Cheese: Teleme

Teleme starts out as a firm, mild white cheese. After a few weeks in the fridge it softens and begins to resemble a Brie. Every recipe I find for Teleme is different. I can't remember where I got this treasured recipe, but I have tweaked it so much that I now consider it mine.

2 gallons whole milk
½ cup prepared mesophilic starter (Available from New England Cheesemaking Supply)
½ teaspoon liquid rennet (or 1 rennet tablet) diluted in ½ cup cool, unchlorinated water
½ cup kosher salt

1. Warm the milk to 86°F. Add the mesophilic starter; stir gently but thoroughly. Cover and set in a warm place. I usually leave it overnight, but an hour will do.
2. Add the rennet. Stir for 2 minutes. Cover and let sit for 30 to 45 minutes, until the curd is firm and a small amount of whey appears on the surface.
3. Cut the curd into ½-inch cubes.
4. Place the pot in a sink full of hot water and slowly heat the curds to 96°F to 100°F (I prefer 96°F), increasing the temperature by about 2 degrees every 5 minutes. Then hold the curds at that temperature until the curds are firm. Stir gently from time to time. To test for firmness, squeeze a small handful gently, then release it quickly. If it breaks apart easily and shows very little tendency to stick together, it is ready. This should be 1½ to 2½ hours after you added the rennet.

5. Pour the curds and whey into a large container lined with wet cheesecloth. Lift the cheesecloth set it into a large colander to drain. (Reserve the whey for another use.) Work the curds with your hands occasionally to keep them separated, and let them cool to 90°F, which may take up to an hour.
6. Line your cheese press with wet cheesecloth. When the curds have cooled to 90°F, spoon them into the press. Insert the follower and press for 1 to 2 hours at moderate pressure.
7. Prepare a brine by dissolving ¼ cup salt in 1 quart warm water. Remove the cheese from the press, cut into cubes, and immerse the cubes in the brine. Let sit for 24 hours.
8. Drain and rinse the cubes, and allow them to dry slightly. Return the cheese to the press and press for 18 hours; this time use firm pressure.
9. Prepare a weak brine by dissolving ¼ cup salt in 2 quarts warm water. Remove the cheese from the press, set it in a deep bowl, and pour the brine over it to cover. Let sit for 3 to 8 days.
10. The cured cheese will be white and creamy. Drain the cheese and contrive a little rack, preferably nonmetal, for it to sit on in a larger container. Cover and store in the refrigerator. Drain the container as necessary.

This is an excellent cheese and quite forgiving as to pressing pressure and length of brining. It takes a bit more trouble at the beginning than does a hard cheese but has the virtue of being ready to eat very soon.

Troubleshooting Off Flavors

Milk and milk products do occasionally have off flavors or poor keeping qualities. Such problems can arise from many variables and sources. Isolating the problem involves asking questions about the entire supply chain. Is it the animals' feed? Is something the matter with the cow, such as low-grade asymptomatic mastitis? Or is the milk handling responsible?

I expect my fresh milk to last at least a week in the family refrigerator. It should last two weeks or even three in the milk refrigerator, which maintains a uniform temperature of 35°F. If it doesn't stay sweet, I ask myself the following questions: Is it being strained and chilled promptly to about 34°F to 40°F? Is the refrigerator staying cold? Are the storage containers clean enough? Is the milk coming in contact with something that we aren't remembering to clean, such as the lip of a pitcher?

Isolate the Problem

To begin, you can try to figure out whether an off flavor is coming from the milk or the milk handling. Check the milk from each quarter of the cow separately for flavor. The easiest method is to milk some from each quarter into a cup and taste it immediately. It should taste delicious. If it tastes salty, this is a sign of mastitis. If it tastes flat and uninteresting, the cow may have some other sort of problem in that quarter. If you are not ready to try milk warm from the cow, get four jars, milk some from each quarter into a jar, label, and chill your sample before trying it.

If the milk from each quarter of the cow seems fine, the problem may be in how you're handling the milk.

Identifying Problems with the Cow

If butter, cream, or milk develops off flavors and the milk taste test described above yields less than perfectly delicious results, check for some of these culprits: Consider your cow's diet and surroundings. Silage, turnips, wild onions, some weeds, and even alfalfa on occasion may impart flavors to the milk that some people won't like. Your own family will probably scarcely notice them, except at first. My children have said the milk was sour at times when the cows had been eating certain foods. Having your cow eat hay for a few hours before milking will help reduce such flavors.

Cows kept all winter on dry feed can become deficient in vitamin E. This lack will manifest itself as a rancid or oxidized flavor in the milk, often described as painty. Vitamin E is usually added to mixed dairy feeds but may be insufficient. Other dietary antioxidants may also be deficient, including selenium and beta-carotene. I like to give my animals a vitamin E supplement in winter to avoid any chance of a deficiency. I make my own supplement by adding the contents of a four-ounce bottle of vitamin E oil to a quart of vegetable oil. I pour a couple of ounces of this on my cow's feed every day in winter. It seems to work. Sprouted wheat is a good source of vitamin E. Kale is the richest readily available plant source of vitamin E for cattle; they love it, and it will stand some freezing.

Some people report that adding a couple of tablespoons of dolomitic limestone to the feed improves the flavor of milk.

Cows kept in foul surroundings will give milk that reflects the odor. Keep the cow's stall clean. (Where have all the farmhands gone?) Some water supplies taste terrible; such water affects the flavor of everything it meets.

Though mastitits is the primary disease causing off flavors, ketosis, a metabolic disorder (see page 81), can also affect the flavor of milk.

In late lactation the milk of some cows changes flavor. Some people detect an altered flavor when the cow is in heat.

Though there is sometimes improvement to be had, and of course illness can be treated, sometimes identifying the problem of off flavors in milk won't help you solve the problem. The dietary causes of off flavors may have to be temporarily endured until management has been altered or the season of a particular weed passes.

Identifying Problems with Milk Handling

If off flavors are due to unsterile equipment or containers coming in contact with the milk (and this is the most common reason), an energetic scrub-down is in order, followed by scalding. Scalding is cheap insurance against loss of dairy products due to off flavors.

Avoid using a sponge for washing up, as it can harbor and redistribute bacteria. A brush is best. I keep a brush exclusively for milk utensils, and I never use it on the outside of buckets. Ideally, you should have a special room for milk handling, something few of us nowadays can manage.

If you have trouble cooling the milk quickly owing to lack of refrigeration, you can make a cooler using running cold water. Set a tinned or stainless-steel milk container (with a cover) inside a larger container, and set water running over the covered milk container in the outer container. The warmest water will run away. Or work on the evaporative cooler principle and drip water down a burlap sack or towel wrapped around a milk can. As a drip hose, run a section of pierced copper pipe or plastic soaker hose in a circle around the can, connected to a slow-running hose.

If you are using a refrigerator to cool the milk but it works slowly, remember that the coolest place is on the bottom shelf. Try to train the family to open the door less often, and avoid putting strong-flavored foods in with uncovered milk.

To achieve rapid chilling some readers tell me that they put their milk jars into the freezer for a predetermined number of minutes. This works fine until the day you forget to remove them.

Off flavors can arise from contamination during the incubation of dairy products. When setting your cream to ripen, cover it lightly with a tea towel. Air circulation is necessary for natural souring (as opposed to deliberate inoculation, as in making yogurt), but milk or cream left to ripen while tightly covered usually tastes bad. On the other hand, if the cream is uncovered, it can easily pick up undesirable odors, such as onions or smoke. It is primarily the fat that picks up odors. Skim milk setting for cottage cheese is less susceptible.

If circumstances force you to use plastic containers, check whether they could be causing off flavors due to something previously stored in them. Plastic is remarkable for its ability to retain odors.

Plastic and Aluminum

A final word about containers is in order. You probably could use aluminum satisfactorily for milk, just for brief contact, if you have to. Although there are health concerns about aluminum, these apply to prolonged contact particularly with acidic food. Sometimes the only strainer you can get will be aluminum. The surface will always be hard to keep clean; it seems to collect scum. If used for acidic foods, it readily becomes pitted and smells metallic and unquestionably dissolves into food.

Recent articles in scientific journals have described potential hazards from plastic. Ingredients in various plastics are soluble in both water and oil. This means that ingredients in plastic can dissolve into both skim milk and cream. That chemicals found in plastic can cause cancer and behave as estrogen analogs is not in doubt. The only question is how much it takes to result in health problems. I prefer not to use plastic containers for milk. Milk stored in plastic picks up a distinctive flavor. In addition, plastic is extremely difficult to sterilize. Reusing plastic milk jugs will result in milk that doesn't keep very well. Plastic buckets are tippy and don't last long. I don't think plastic deserves much of a place around the home or farm.

CHAPTER 5

Drying Off and Milk Fever Prevention Diet

The end of lactation calls for a succession of dietary changes intended, first, to discourage milk production, then to build up the cow's nutritional reserves and support the unborn calf, and finally to protect her against the metabolic imbalance that can lead to the condition known as milk fever.

Drying Off

Before your cow has a new calf, she should be dried off, which means helping her to stop producing milk. Once you are absolutely certain she has been bred (it's best to get a vet check), mark her due date on your calendar. She should have a dry period of fifty to sixty-nine days. Count back from her due date and mark the date to begin drying off. If when this date arrives she is giving twenty pounds or less (two and a half gallons) of milk each day, simply stop milking. Cut out her grain (this does not affect rumination) and feed her your second-rate hay. Don't bring her into her milking area or do any of the things that encourage letdown. Check her udder twice a day. Allow it to get full but not hot. If she seems uncomfortable, bring her in and milk her out completely; give the milk to animals, as it won't taste right. Continue low feeding and checking her udder. When her udder fills again it will likely not get hard. Unless it worries you (because it has a hard spot, for example), leave it alone. Within a week to ten days it should stay flabby. Don't test it by taking a few squirts, as the teats will have formed an antibacterial plug that should be left in place.

If she is giving more than twenty pounds of milk a day when you are ready to dry her off, put her on low feed and milk her once a day for three days, then skip a day, milk once more, and start the above schedule. Use your own best judgment and adjust these suggestions to circumstances. When you milk her out it is best done completely; leaving her milked halfway is an invitation to mastitis.

Following the last milking, most authorities recommend a dry cow treatment, which involves an antibiotic inserted into the teat canal and an antiseptic teat dip, which leaves a protective coating. These are preventive measures against mastitis. Mastitis can sneak up on a cow during the dry period or prior to calving when the udder gets full and drips. These measures are routine in dairies. I am unable to judge their importance for a family cow. I have never done it. The one time I encountered fresh cow mastitis, a lapse in management (mine) was the cause; I had delayed milking her out. But if mastitis has been a persistent problem or if your cow's surroundings are very muddy, a dry cow treatment may be advisable.

Instruct everyone not to attempt random squirts of milk at any time up to actual calving. You absolutely do not want the teat end opened and the waxy plug lost.

Warning repeated: Before drying off your cow, be sure to establish with absolute certainly that she is in calf. You may need a vet check or wish to do BioTracking, a pregnancy blood test for ruminants (see the resources). If you dry off an open cow – and it does happen – it may then be well over a year until you have milk again. A long vacation from milking while not in calf doesn't do a cow any favors health-wise. She is almost certain to get too fat, and hormonal disturbances may also occur.

Evaluate Her Condition

It is especially important to evaluate your cow's condition at the beginning of drying off. Weight can be estimated by measuring with an inexpensive dairy weight tape available from feed and agricultural supply stores. To use the tape, measure her girth just behind her front legs. The tape will be marked with weights matched to girth for the three major dairy breeds, Holstein, Guernsey, and Jersey.

A more accurate estimate takes account of body length. Take the heart girth measurement (in inches) and the body length measurement (point of shoulder to pin), then perform this calculation: heart girth × heart girth × body length, then divide by three hundred. The result is a good estimate of your cow's weight.

Expect to be able to discern the first three ribs of a dairy cow. A fatter cow is storing extra abdominal fat, which may interfere with calving. A thinner cow will struggle to support herself, her calf, and her milk production, and something will be compromised.

As soon as you are satisfied that your cow is dried off, put her back on good hay. If you decide to use anionic salts (see page 83), continue regular grain feeding. Or if she has been on pasture only, merge in any high-calorie grain or premixed ration you wish. If at any time she is too fat, slow down on everything but hay or grazing. She will still need the volume and quality of hay or grazing, but the extra carbohydrates of grain can be limited. You can stop grain altogether, but you will then have to find some other way to feed the anionic salts. As we'll shortly discuss, milk fever prevention diets call for a tricky balance, which explains why much continues to be written on the subject. And as an aside, note that milk fever is seen less often in the grass-fed herd.

Importance of the Dry Period

Except for the week or so when you are drying off your cow, her feed may need to be changed but not reduced. During her dry period a cow is making a new calf and storing nutrients for her next lactation. If her previous lactation diet included a significant amount of grain, she will profit from a chance to rest and restore her rumen. A cow's stomach ferments feeds other than hay at a lower pH. This more acidic environment induced by grain can erode her rumen. A rest on a diet of mostly hay and grazing will give her rumen a chance to rebuild its surface and its muscle tone, which inevitably declines when long fiber has been restricted. However, avoid abrupt changes of diet. (For a discussion of rumen function, see chapter 9, "Feeding Your Cow.")

If your cow is very thin, this advice does not apply. The need to gain weight during the dry period overrides other considerations, at least temporarily.

Inattention to diet during the dry period not only prejudices a cow's ability to milk well next time and have a strong calf but will set her up for milk fever and ketosis. These are metabolic disorders, or hormonal derangements if you will.

Milk Fever and Ketosis

Milk fever has plagued dairy cattle from antiquity. It also strikes goats and is not unknown in dogs, cats, and other species. It is a form of paralysis resulting from failure to mobilize minerals, especially calcium, fast enough to meet the sudden demands of milk production. Calcium levels in blood are critical and must be maintained within a small margin of error, much as is the case with oxygen. Calcium is essential to muscle function. If calcium levels suddenly drop due to the imperious demands of the udder at parturition, muscles cease to function and paralysis results. In general our bodies, and those of cows, are so good at maintaining correct blood calcium levels that we don't have to worry about it. But at calving time the metabolic balance may be overwhelmed, especially for heavy-producing cows.

Milk fever affects about 8 percent of cows at calving, with Jerseys being more susceptible due to their greater ratio of milk production to body weight. Before studies clarified the cause during the 1950s, some bizarre folk remedies were applied. These included cutting off the cow's tail, insufflating the udder with air or water, and worse. It is now understood that dietary management during the dry period is the key to preventing milk fever. Symptoms and emergency treatment are discussed in chapter 15, "Treating Milk Fever." Be sure to read this section, and be prepared to call the vet or treat the problem yourself should milk fever occur. It is an emergency and without treatment is fatal more often than not.

Ketosis is a metabolic disorder that may affect a cow after she recovers from milk fever, though it may also come on without this preamble. It results from rapid mobilization of fat stores in the cow's attempt to meet the energy demands of lactation, which causes an abnormal buildup of ketones in the bloodstream. Ketosis can become chronic or even fatal. Loss of appetite, especially for grain, is the chief symptom, along with a sharp reduction in milk production. The other name for ketosis is acetonemia. The cow's breath smells of acetone, like what is found in nail polish remover. Some people describe the smell as more like overripe pears. It isn't an emergency like milk fever but requires treatment.

The Dry Cow Diet

For many years a low-calcium diet was standard practice for a dry cow. The intent is to keep the hormones that mobilize calcium on standby alert and ready for their huge job when the cow freshens. A more recent type of dietary management requires restricting potassium so that the blood remains slightly acidic. Phosphorus intake must also be controlled because it not only favors alkaline conditions but above a certain level will suppress the parathyroid hormone, which must be active in order to synthesize vitamin D, which is key to calcium utilization. Nowadays one common approach to feeding the dry cow calls for using anionic salts to support calcium mobilization by maintaining an acidic balance in the blood.

The Low-Calcium Diet

Low-calcium feeds include corn in the ear, corn grain, corn silage, barley, and barley straw. Locating these feeds and restocking your cow's pantry with a special diet can be a deterrent to their use. And changing feeds is a challenge for dairies of any size. A simpler and arguably more effective approach employing anionic salts is now generally favored. It does not require restriction of calcium.

Anionic Salts

All soluble minerals have an electrical charge, either positive (cation) or negative (anion). Potassium, sodium, calcium, and magnesium hold a positive charge and are cations. Chloride, sulfur, and phosphorus hold a negative charge and are anions.

The balance of minerals in a feed determines its dietary cation/anion difference (DCAD). Those feeds with a higher ratio of cations result in alkaline conditions in the bloodstream (higher pH). Those with a higher ratio of anions result in acidic blood (lower pH). Alkaline rations have been found to predispose to milk fever, whereas acidic rations help prevent it, in part by aiding the release of calcium from bones and the gut. The problem here becomes one of increasing acidic feeds to aid mobilization of calcium while limiting high-potassium feeds because potassium is a cation. Since this is just as difficult as putting together a low-calcium diet, formulations of anionic salts are available to add to feed to boost acidity.

Some feed stores sell anionic salt mixtures with suggested rates of feeding. These can be mixed with anything palatable. If you use them, experiment and find out what works. Start adding the salts ten days to two weeks before calving, not sooner.

The addition of anionic salts permits more latitude in feed selection because the salts promote acidic conditions in the bloodstream irrespective of diet. The point is not so much to enable calcium uptake, which it is now believed will take care of itself in even mildly anionic conditions, as to overcome the cation influence of potassium. If it is possible to feed your cow corn silage, a low-potassium feed, a lesser amount of anionic salts will be necessary in order to reach the desired pH levels in the bloodstream.

Anionic salt mixtures are not palatable but apparently taste even worse when premixed with grain and allowed to stand. The salts are better accepted when freshly mixed into feed. Those that I have purchased work out to ¼ cup of the actual salts on feed twice a day. Premixed and pelleted salts are available.

Do not overdo it with anionic salts and push the DCAD ratio too far or kidney damage can result. Graduated paper test strips are available to check the acidity of a cow's urine. Test about four hours after she eats her dose. She will oblige with a urine sample if you stroke next to the vulva. Normal urine pH is about 8.0. With successful DCAD treatment, Jersey urine will be 5.5 to 6.0 and that of larger breeds 6.2 to 6.8.

A convenient treatment option that has received favorable mention from the scientists at *Hoard's Dairyman* magazine is a pelleted anionic preparation called Animate (see the resources at the end of this book). It is palatable and calls for no accompanying feed changes. I have not yet tried it but plan to when next my cow calves.

FEEDING CALCIUM

Dietary calcium must not be restricted too radically. Suitable feeds with moderate calcium content include rye, oat, and timothy hay or grass, which is convenient information for graziers. Alfalfa at all stages is one of the feeds highest in calcium and probably best not fed until after calving. Needless to say, protein cannot be ignored. For those who would like detailed analyses and formulae, a good source is *Caring for Transition Cows* by Mike Hutjens and Earl Aalseth, available from Hoard's Dairyman (see the bibliography).

Calcium gel (150 grams) should be given on the day before and the day after calving and high-calcium feed provided from this day forward.

The Vitamin D Option

Another measure now recognized as effective in preventing milk fever is the massive vitamin D shot. This is a stand-alone treatment not combined with diet changes or anionic salts. A single injection of ten million IU of crystalline vitamin D (IV or SC) given eight days before calving is likely to be effective. It can be repeated if a cow doesn't calve on time.

Other Considerations

If there were simple final answers, less discussion would be needed and research could cease. In general, if left to chance, a first-calf heifer is least likely to develop symptoms. Any cow that was previously stricken has a greatly elevated chance of a repeat of both milk fever and ketosis. (Of course, if your cow was previously owned the former owners will probably play down any such problems.)

A cow on good pasture that includes a variety of grass species and forbs (broad-leafed "weeds") may balance her own nutrients in such a way as to preserve herself from milk fever. In these circumstances the rumen flora become more efficient at extracting nutrients. Evidence for enhanced efficiency is found in the decrease in methane production by cows on good mixed pasture. Methane (CH_4 or natural gas) is essentially the same high-energy gas that we use in cooking. Methane that the cow burps up (eructation) is energy lost to the cow for her own purposes. Nonetheless, eructation is normal, and some will always occur in all ruminants.

An additional factor undoubtedly working to prevent milk fever in cows on pasture is vitamin D. Calcium is not absorbed or processed without vitamin D. A cow can make her own through her skin if she is outdoors in the sunshine. Vitamin D is also formed on sun-cured hay. Obviously these factors vary according to circumstances. Calcium uptake and mobilization are also dependent upon function of the thyroid and parathyroid glands, as noted, and these in turn must be supplied with adequate iodine. Kelp is a reliable source of iodine. I make a practice of adding a handful every day to my cow's grain. It can also be offered free choice from a fixed box in your cow's loafing area.

Milk fever is unpredictable. One year my cow Helen nearly died of it on Easter. She was saved only because my vet came straight to the farm despite the holiday and administered calcium IV. The following year I went out on a sunny early morning in early June and found a big healthy heifer standing at Helen's side among the buttercups, and she wasn't sick a day. Who knows why? Perhaps because it was two months later in the spring and the combination of sun and good grazing made the difference.

Some dairymen say they never see milk fever. They usually have totally grass-fed herds with lower production, with many years of fine-tuning the management. I applaud their success. For family cow owners, often with a new cow and unimproved pasture, perhaps having to settle for whatever hay is available, I suggest that it is best not to be overconfident where milk fever is concerned. Institute a feeding plan, make emergency preparations, and don't feel like a failure if milk fever occurs despite your best efforts. Given treatment, most cows make the miracle recovery described in the section on treatment (see page 218).

My grandmother used to quote old-timers as saying milk fever hit "always my best cow." We are fortunate now to have informed approaches to this formerly often fatal disorder, which does indeed disproportionately affect the best producer. The last word has not yet been written on milk fever, and research continues. I can think of few more useful applications of tax dollars than the research on milk fever, much of it done by our land-grant colleges.

CHAPTER 6

Calving

Your cow will in most cases give birth without assistance. All you need do is be available in case you are needed. It is much better not to interfere with calving unless you observe signs of trouble. But you can stay with your cow in labor and watch the birth. Because she is used to you and your family, your presence usually will not worry her. The presence of strangers will slow down labor.

When you suspect that calving is near, clean your cow's stall and put down a heavy mat of dry bedding. Have water and hay available, and bring her in to calve. She can calve outdoors just as well, if the weather is mild. But part of the enjoyment of keeping a cow and compensation for all the work is being present and participating in the big event of the birth of a calf.

There are two other practical reasons for calving in the stall instead of outdoors. One is the possibility of the calf wandering off or falling over an embankment or otherwise becoming separated from its mother. The other reason is that there is always the possibility of milk fever, and you'll feel better if you can keep an eye on your cow during and after calving. Expect everything to go easily, but take precautions nonetheless.

Gestation in cows averages 279 to 290 days, with Jerseys having the shorter gestation.

Signs of Calving

Our veterinarian used to tell us, "I can tell you anything you want to know, except when a cow is going to calve." In general your cow will calve a little later than you expect, unless she calves suddenly a little early. There are some clues besides the mark on the calendar. As her ninth month arrives, viewed directly from the back or front she will appear round, like an orange on sticks. When calving is imminent the calf shifts position and aligns itself with the birth canal; all of a sudden Bossy looks slab-sided. Signs you can observe on the cow's body are the rapid swelling of the udder (in contrast to the gradual enlargement that has been taking place) and a spreading of the space between the base of the tail and the pinbones on either side of it. You can put your hand on the space between the tail and pinbone and feel hollowness and wobbliness of her tailhead that wasn't there before. (You should feel the relationship of these bones well before her due date so that you have a basis for comparison.) The wobbly tail head means that pelvic ligaments are softening. Generally when this opening of the birth canal occurs, labor will begin within a day. The teats get swollen at the top, giving them a conelike shape. There will usually have been strings of mucus from the vulva for several days, particularly in an older cow. When the mucus becomes blood tinged, this is a definitive sign.

Behavioral signs of imminent calving are the cow turning and looking at her back end, kicking at her belly, restlessly moving in semicircles, stepping forward and back, and having a straddle-legged stance. She will stop cudding. She may get up and lie down repeatedly. If she is leaking milk and notices a puddle of it, she may paw at it and give the soft murmuring calf call.

When a large, fluid-filled membrane begins to emerge, you know for certain calving is under way, for this is the water bag in which the calf has developed. By the time you observe the water bag your cow will have been in labor for several hours. A heifer calving for the first time will often take six hours to get to this point. During this first stage, the cow may or may not eat fitfully. A cow will sometimes eagerly seize a mouthful of hay and then lose interest before eating it. She may show signs of irritation and shake her head at you.

In the second and more intense stage of labor, your cow will have less time for you, but I do think that my cows have at times appeared appreciative of my company and support. A cow can calve either standing or lying down, with standing more common. As the water bag moves through the cervix and the vagina, the feet should appear. Watch for something white, which will be the edge of a hoof. This is likely to show and then recede between strains. Hopefully two right-side-up front feet appear. In the correct and most common presentation of the calf, the front feet will be just ahead of the nose, which rests between the calf's front legs. (We'll talk about abnormal presentations below.) When you see two front feet and a nose, you know the calf is in the correct position. The water bag usually ruptures at about this time, spilling out a flood. Sometimes you can see the shapes of feet and nose within the bag before it breaks. Don't be alarmed if the calf's tongue is hanging out.

After the feet appear, the intervals between bouts of straining become short, ranging from fifteen seconds to two minutes. The strains increase in intensity and last a few seconds each. There will be rest periods between groups of strains, often just after the feet and nose appear and again when the head is out, and from there the rest is relatively easy. As the chest comes through, there will be more fluid as the remaining contents of the water bag are expelled. Mucus will pour from the calf's mouth and nostrils, opening these passages and preparing them to take over the respiratory function. Occasionally a very large calf will get stuck at the hips and you will need to rotate it. A few more strains and the calf is born. This second stage of labor may last two hours but is often surprisingly quick, giving you barely time to run back for your camera. A heifer takes longer than an older cow, and a large calf will cause the process to take longer as well.

During a slow delivery you can give valuable help by grasping the hoof and pulling down while the cow strains. Don't pull the leg out past the knee joint unless a second hoof appears. When a second hoof appears, grasp the two of them and pull *down* while she strains. Now the calf's nose should appear. This is proof that the calf is lying correctly and you can safely continue to help. Take hold of the slippery forelegs, using a towel for better grip, and with her next contraction pull down hard. A few pulls while the cow is straining will deliver the calf if there are no complications except fatigue.

Advice from the research team at Ohio State University (published in the November 2011 *Journal of Dairy Science*) is to begin assistance seventy minutes after the amniotic sac appears or sixty-five minutes after feet appear outside of the vulva.

When the Calf Is on the Ground

Check the calf at once to be sure it does not have part of the sac around its nose. If the calf does not appear to be breathing, pick it up by the hind feet and hold it vertically for a few seconds (not more than three or four seconds). This will usually start respiration. It's normal for the calf to cough and snort with its head flopping around; if it still isn't breathing after you've hoisted it, try tickling inside its nostril with a piece of hay.

The cow will immediately turn around to start licking her calf. If she is down and doesn't rise at once, pull the calf around to her head. She will then greet her offspring by starting the thorough licking job that dries and warms the calf. The cow will identify her calf by its distinctive odor as she does this, and no other calf will then suit her so well. A cow bonds intensely with her calf.

Within fifteen minutes the calf usually starts staggering to its feet, rear legs first in the way of cattle. The initial attempts invariably result in a series of nose landings. A calf usually requires a half hour of effort to get it all together and stand on four feet, wavering. A step toward the cow usually results in collapse and a new start, but with each try the calf becomes better coordinated.

The calf should have colostrum (the first milk after calving) as soon as possible. A calf is born without antibodies to disease, so it badly needs those antibodies that are present in the colostrum. To permit their rapid passage into its bloodstream, the calf's gut is permeable. If it gets bacteria in its mouth well in advance of colostrum, these have a royal road right into the bloodstream. Colostrum coats the gut and protects it; closure occurs quickly after the antibodies have crossed the gut wall. After this, colostrum continues to be valuable for its high vitamin, fat, and protein content, but further antibodies pass more slowly. If colostrum is not obtained, the gut wall remains open, waiting for it, for some days, during which time the calf remains highly susceptible to infection. Virtually no calves survive without colostrum. If you should have the misfortune to lose your cow at calving, milk out some colostrum while you can. Normally the cow has ten times what the calf needs and I always keep some frozen just in case. It can also be used to save orphans of other species.

The final stage of labor is pushing out the afterbirth, or the placenta. This often occurs while the sucking is getting started. The placenta appears slimy and stringy. It is worth taking the trouble to know for certain that it has been ejected. A cow will usually eat the placenta, or a large portion of it, unless it becomes contaminated by dung. Some cow owners whisk away the placenta on the theory that the cow might choke on it. However, there is considerable evidence that the cow benefits from the hormones and nutrients in the placenta, which assist in contracting the uterus. I always allow my cows to eat it, but I remove whatever is left as soon as interest is lost.

Often strings of placenta remain dangling from the cow. Resist the impulse to pull on them, as this may damage the uterus. Also resist the impulse to cut them off. The dangling weight helps the placenta to separate, and if you cut it off some of it may slither back into the uterus, introducing infection.

If you don't find the placenta, the cow probably ate it, but keep checking. If it is retained it can cause severe illness. If you don't see any sign of it after twenty-four hours following calving, you may wish to inform your veterinarian, who can then plan to do something about it. Some veterinarians give a hormone injection to encourage further contractions to expel the afterbirth. Others may wait a few days and then remove it by reaching into the uterus with a gloved hand to find whatever is there; they may implant an antibiotic pellet to control any infection that may have been introduced.

When to Help

If you decide to call a vet to come help with the birth, and you are not well acquainted, explain to him or her that this is a precious family cow. Many have grown blasé from working with big dairies. In the meantime, prepare the site for an emergency C-section, should it be necessary. A couple of bales of hay will do, spread out, preferably where the light is good. A good place for a C-section is a livestock trailer or a horse trailer with the divider removed. If you have not already done so, offer your cow an energy drink of a cup of molasses in a bucket of warm water. It may be useful to walk her around, if you can do so without having her escape to a distant spot. If there are strangers or a dog, ask them to withdraw for the present. Factors that might otherwise not disturb your cow may be making her nervous now.

The most difficult thing about calving a cow is keeping out of it and letting the cow do it on her own while at the same time knowing when it is time to help. Your cow may appear to be struggling so much in pushing the calf's head and feet through a small opening that you feel you must assist. But it is imperative that no pulling from the outside be done until the head has cleared the cervix (the opening of the uterus). The cervix dilates as the water bag and head push against it, and the process must take its time and work little by little, not by brute force. After the head is clear of the cervix, it will have another tight squeeze at the vulva (the vaginal opening). If the cow appears exhausted at this point and the calf appears to be large, it can be helpful to pull on the calf's feet to assist the head in clearing the vulva while the shoulders are coming through the cervix, as described above. What is surprising is just how much pulling is required to do any good. The cow is already pushing with great power, and a little human pull doesn't add much.

If you must pull, it may be necessary to fasten ropes around the calf's ankles and then wrap the ropes around your wrists or have them tied on sticks to make T handles. There are ropes and chains made for this purpose. Good calving rope is of soft material, three-eighths of an inch in diameter and six feet long, with a loop made at each end, and washed clean, and it is a good thing to have it on hand. The pull must be downward, toward the cow's udder or back feet at something like a forty-five-degree angle – *do not pull straight out*. Pulling downward moves the calf through the vulva without tearing. Pull when the cow strains, hold what has been gained when she relaxes, and then pull again when the next contraction starts.

If the cow stops having contractions due to milk fever, it will be necessary to pull the calf out without the mutual effort of the contractions. Great strength is required. Now you *have* to see the calf's nose. The nose should lie between the calf's feet, proving that the head is not doubled back. Since the umbilical cord could become tangled or crushed and cease to provide life support, it is important that the calf be able to breathe without delay. Just because the nose is out, don't think the calf can breathe. It must have its lungs out in the open in order to expand and take in air. Pulling at this point isn't dangerous, since the largest part of the calf is already through the cervix. But the pull must be at the downward angle to protect the vaginal tissues.

A simple rule for deciding whether to assist is that if the cow is making any progress at all, however slow, no pulling is necessary or advisable. But your own judgment must prevail.

The umbilical cord will break when the calf emerges. If not then, it will break when the cow gets up. There is no hurry about this. The pumping through of oxygenated blood may have value. The cord can be cut, dipped in iodine, and tied with a string. While it's not absolutely necessary, it is advisable to do this when there are numerous animals and wet, muddy conditions. Navel ill is a terrible thing, so better safe than sorry.

Abnormal Presentations

I hope you will never need to use the information in this section. It is far better to have a veterinarian or experienced farmer handle an abnormal presentation of the calf. In isolated districts, sometimes that is not possible. Sometimes the only vet is a small animal doctor; in this case he or she may value your suggestions.

The calf should present its front feet first, followed immediately by the nose. In any other presentation, help may be needed to shift the calf inside the uterus so that it can be born. If you are without help and must examine the cow yourself, there are precautions to be taken. The cow will probably view your help with apprehension. You need to have her securely tied so she doesn't move all over the stall while you are reaching inside. It is important that she does not slip and fall. If your stall has a concrete floor, remove the bedding from under her and cover the floor with sand or grit. If the stall has a dirt floor, a bed of hay, straw, or sawdust will give her a good footing.

Scrub your arms thoroughly with soap, not detergent. Wash the vulva with lots of soap and warm water. Then use soapsuds as a lubricant for your hand and arm as you reach in. Reach into the vulva, on through the vagina, and into the cervix to determine the position of the calf. Take your time and make sense of everything you feel as you go. If the calf's nose and feet are there at the mouth of the uterus, the presentation is normal and no action should be required. If the nose isn't right there lying on the hooves, try to determine just how the calf is positioned.

In the breech presentation, the calf's tail would come first if it could, but with the back legs folded under the body the mass is too great for the birth canal. This is corrected by pushing the calf away from the cervix and then pulling the legs, one at a time, upward into position to be born back feet first. While this is a fairly simple procedure, great care must be taken to avoid tearing the uterine wall with the edges of the hoofs. You will need to pass a rope with a noose on the end under the calf, between its legs. Loop the noose over the first hoof you find and pull the noose tight on the pastern (just above the hoof). Use one hand inside the cow to guide the hoof and protect the uterine wall, and the other to pull on the rope to bring the hoof up to the cervix. Helpers who can calm and steady the cow and help manage the rope are valuable here. Use a second rope to repeat the process with the other hoof and leg, and the birth should then proceed normally. If the cow is exhausted from a long laboring effort to deliver the undeliverable, you can help her with the ropes attached to the feet.

A calf with its hind legs coming first can be born without assistance provided the front legs aren't also coming at the same time. These may need to be pushed back until the rump is in the birth canal and the calf is uncurled and straightened out.

If the calf is presenting in the forward position but its head is turned back, it will be impossible for the calf to fit through the birth canal, and you'll have to pull the nose around to the forward position. Reach into the uterus and push the calf back, away from the cervix. You will need to do this between labor strains that obviously will be exerting the opposite force. Then stand with your back to the cow, reach your arm back into the uterus, and work the head of the calf into the forward position. You will need to stand backward because the human arm will not otherwise bend in the direction required to accomplish this maneuver in the space available. It takes considerable strength.

If the calf's nose is forward but its front legs are back, it is again necessary to push the calf back between contractions and then bring the leg or legs up, cupping your hand over each hoof to avoid tearing the vaginal wall. This is difficult because of the great pressure forward.

In the rare case of twin calves, one can jam against the other. When it is necessary to explore the birth canal with a lubricated hand, it is useful to bear this contingency in mind. The "lead" twin can be released by pushing the other back, and then delivery can proceed sequentially (and normally).

Any presentation that requires twisting or moving the calf presents a danger of pinching or breaking the umbilical cord. Once interference has begun, it is vital for the calf that the birth be accomplished quickly. If you decide you must take on a birthing problem yourself, be prepared to see it through. You should get an antiseptic bolus (pellet) to place in the uterus after calving to combat any infection you may have introduced. If possible, however, get a veterinarian, and then learn all you can by watching and assisting.

The calf must come out or you will lose both cow and calf. But be assured, these problems are rare. Most common is that you go out to check your cow and she and the calf are standing there looking happy.

Photograph courtesy of Max Luick

CHAPTER 7

The Calf

Of the many rewards of cow ownership, not least is the gift each year of a calf. The calf is of great value whether it is a bull or a heifer, albeit for different reasons. Early care is the same irrespective of its sex.

Colostrum as Soon as Possible

The calf must have colostrum as soon as possible, ideally within thirty to forty-five minutes of its birth. All studies confirm that the sooner it gets colostrum, the better its chances of survival. A calf born in spring will often be on its feet within that time. It will often need help finding the teat, since it is preprogrammed to seek the teat up where it used to be prior to selective breeding, which was a lot higher off the ground. The calf should have a gallon of colostrum within the first twenty-four hours.

Helping the Calf to Suck

In case the calf proves inefficient in finding the teat, you can help it, though the job will be much easier if a second person is available to assist. Tie the cow or have someone hold her head. Otherwise she will keep circling to lick and nudge the calf and it will keep tipping over. The instant the calf gets the teat in its mouth it begins to pull back and usually falls over, especially if the cow is licking it. Pushing the calf into range of the teat works against its instinct to pull back. You can't pull it forward either, because it sets its feet. Nudge it forward instead. Once sucking properly, the calf will alternately pull back and bunt with surprising vigor. The word *exuberance* comes from the Latin word for udder, and its root meaning is "with the enthusiasm of a suckling calf." You will see.

If the cow is down and will not get up, milk as much colostrum as you can into a bottle from whichever teats you can get at and bottle-feed some colostrum to the calf. (You should have one or more calf bottles on hand even if you plan to have the calf suckle the cow.) A calf should be fed standing on its feet. If you are working alone, back the calf into a corner to prevent it from backing away from you, and straddle it. The calf's neck should be down but its nose tipped up. Its body should be in line with the neck, and not twisted. There is a structure in the calf's throat, called the milk groove, that sends the milk past the immature rumen and into the true stomach. The milk groove doesn't function properly unless the calf is in the correct position.

A weak calf, usually a winter calf, may be hard to get sucking properly. Its tongue will loll out and it will flop over. If it cannot stand, then hold it in a kneeling position, supporting it evenly from both sides, with its front legs folded under and its neck and head positioned as described. Otherwise the milk will go to the wrong place and the calf can even drown. Make every effort to get the calf on its feet before resorting to feeding it lying down. Often it just needs rubbing and warming.

Even if the first colostrum feeding must be from a bottle, once the calf is bouncy and the cow is on her feet it is easy to encourage suckling. Just squirt a little milk on the calf's nose from point-blank range so it sees where milk comes from and leave them together. The cow will assist.

A word of warning: Don't assume that because the calf is standing next to the cow nuzzling around that it has actually latched on and sucked. You need to observe it feeding. Even if it seems to have the teat in its mouth, wait until you see its tail wagging and head bobbing and foam accumulating. If you arrived late on the scene and are not sure if the calf has fed, look at the hair around its nose. It should be shiny with dried milk. Often there are puffs of foam on the ground. The flanks of a newborn are so caved in that it almost seems you could shake hands with yourself under its ribs. After feeding, it fills out. If you are really concerned, you can milk out some colostrum and offer a bottle. Keep an eye on the situation. I have known of tragedies where it was incorrectly assumed that the calf was feeding based on the fact that it was lively and was seen to nuzzle the udder. Like most babies, a calf is born with a couple of days' reserve energy, after which it will collapse if it does not receive nourishment.

If you leave the cow and newborn calf together, make very frequent observations. If there is the least suspicion the cow is developing milk fever, separate them at once. A cow will take excellent care of her calf and never overlay it unless she has milk fever. The resulting paralysis will render her unable to control herself.

If you leave the calf in with the cow to suckle at will, it will often choose one or two teats and neglect the others. This may lead to mastitis. You need to keep the milk moving, and the calf is your best aid. Try to get the calf to suck the neglected teat(s) by pushing it up close and squirting some milk in its mouth. You may need to hold your hand over the other teats to hide them. If the calf refuses, milk out that quarter by hand.

How Much Milk Does a Calf Need?

Most calf care manuals suggest that one gallon of milk a day divided into two feeds is all that a calf needs or should have. I have not followed this advice, and current recommendations are to feed more. A study reported in *Hoard's Dairyman* (August 10, 2003) found superior growth and health on the following regimen developed at Cornell by Michael Van Armburgh:

At birth: 3 to 4 quarts of colostrum
Week 1: 2 quarts two times daily, all fed by nipple bottle
Week 2: 3 quarts two times daily
Week 3: 4 quarts two times daily
Week 4: 5 quarts two times daily

Following each feed, the calves in the Cornell study were offered as much warm water as they wanted, also by nipple bottle.

Calves on this regimen, which also included solid food, gained two to three pounds per day and, compared to controls on the standard one-gallon-per-day feed, had a more vigorous immune system, had fewer respiratory problems, and were livelier. "Occasional runny manure was not due to scours and soon righted itself," the article stated.

My own practice with a calf separated from its mother is to start off in the early days offering half a gallon three times daily. I increase this one quart at a time as the calf seems to want, often by adding a fourth feed late in the evening. By the end of the second week I drop back to three feeds, for convenience and because I have increased the volume of milk at the other feeds. I have not fed any warm water. Neither do I encourage solid food.

A calf has an impressive sucking instinct and will soon be sucking everything in sight if underfed or even following what you might have thought to be an adequate feed. If the calf is with its mother, though, you never see this frantic extra sucking. Nature's plan always includes sucking beyond satiety;

it ensures continued milk production, and as noted above, I don't find that it leads to overfeeding.

Left running with its mother, a calf gains two to three pounds per day (equal to the calves in the Cornell study) on milk alone plus whatever bit of grazing or hay they manage to eat.

Advantages to Separating Cow and Calf

There is a great deal to be said for separating the calf from the cow and feeding it entirely by bottle right from the beginning. Many commercial dairies tie the calf in front of the cow, where she can see and lick it but not suckle it. After three days it can be placed entirely out of sight, or sold, and you can proceed with quiet milking.

A calf as young as three weeks old can be a great nuisance to control if your management scheme involves letting it in and out for feeding. Some people report success with milking the cow partially, leaving some behind for the calf. Every cow I have owned has quickly gained control of her letdown reflex and, knowing the calf is waiting, would not let me have any milk. Or I might get half a pail, but there wouldn't be much cream in it. The cream is in the hind milk and is held back. I've also tried letting the calf in first so that the cow lets down, then dragged it away when I think it's had enough. Soon the calf will be too strong and rebellious to make this a fun part of milking. After you've removed it, it will moo and the cow will moo back, and she may even express her resentment toward separation by kicking and certainly by raising her tail and letting drop.

If you decide to bottle-feed the calf, its surroundings must be kept scrupulously clean and dry. The bottles and nipples must be sterilized after each use. They can be put in the dishwasher if you have one. Don't allow people to amuse themselves by letting the calf suck their fingers. Calves are very susceptible to gut infections of all types.

A calf needs plenty of handling. Rub its back every time you feed it to make it feel loved and to stimulate digestion and elimination. As it learns to take its bottle efficiently, you can stand in front of it when feeding. But don't brace the bottle against yourself; instead, hold it to one side, and train children not to stand directly facing the calf while it takes its bottle. When it bunts it is surprisingly strong and might alarm or even injure a child. If you don't want to hold the bottle yourself, agricultural supply stores offer wire racks in which the bottle can be set.

After a few weeks you can switch to bucket feeding. But digestion of the milk will still be optimized if the calf swallows from its natural sucking position: neck lowered, nose tipped up. There are teat buckets available; they are very efficient for feeding the older calf, especially if you wish to feed more than a half gallon, but they are harder to keep sanitary. The teat bucket is preferable to an open bucket because it assures that the calf sucks rather than drinks, so that the milk goes to the abomasum rather than the rumen, where it does not belong.

Whether using a calf bottle or teat bucket, be particular that the milk is heated to within 100°F to 105°F for proper digestion and reduced stress to the calf. And discard any flabby nipples. Hard sucking produces more saliva, which is important to digestion. The calf needs to work for its living!

The advantages to bottle-feeding a calf lie in greatly simplified management later on. Bottle feeding prevents formation of a strong cow-calf bond. But even seeing their calf is sufficient to keep most cows interested for a while, especially if it isn't their first calf. Total separation for many weeks may be necessary. If grazed together too soon, a cow will stand so as to offer her teat, and it's a dumb calf that will not soon figure out where milk really comes from.

Reliable weaning of a heifer you wish to keep will sometimes prove impossible, leaving you no alternative but to sell her. Strict bottle feeding avoids this outcome.

The advantages to bottle feeding also include a greater and more consistent milk supply available for the house, which is the reason, after all, for having a cow. If you have uses for all the fresh milk, you can feed the calf a milk replacer formula. For superior growth, integrity of bone, and general appearance use a milk-based replacer, not a soy-based one; despite claims, soy-based replacers are inferior. Buy a high-quality product that is 29 percent protein and 15 percent fat. Commercial dairymen use fermented colostrum or a milk-based formula for feeding their replacement heifers, reserving soy-based formulas for raising steers or calves meant for somebody else.

Advantages to Keeping Cow and Calf Together

There are a number of advantages to keeping the calf in a small pen where the cow can see and lick it. After you are sure the cow is past the risk of milk fever, she and the calf can spend the night together. If you put them together following the evening milking, when the udder is empty, a young calf will still get plenty of milk and you will still find milk in the morning. A young calf will usually concentrate on one or two teats. An older calf will have it all. For many people, avoiding one milking is a huge advantage. And the cow and calf do look awfully cute and blissfully happy together.

Earlier calf-care manuals warn of the danger of allowing the calf too much milk. I have never actually had this happen. I suspect that it is separation anxiety combined with hunger that causes a calf to overeat. If left with the cow all the time, it appears to eat just what it needs.

A fresh cow produces at least ten times what a baby calf wants. Expect to milk twice a day at first in any case. In a couple of months, or even sooner, you may go to once-a-day milking. Be sure to check the udder even if you skip milking. Unmilked, the full quarters will surely develop mastitis.

Check daily for cuts on the teats. A calf's lower incisors are razor sharp. Any cut from its teeth is likely to be vertical. A small cut doesn't seem to bother the cow, but a bad cut will cause her to kick the calf off that quarter; then you will have to milk that quarter by hand.

In case of cut teats, I recommend a tool called a Dremel; it is like a small electric drill or a large dentist's drill and has a choice of bits. The carborundum bit works like a charm to file down those incisors. If cuts are a chronic problem, note that a calf getting plenty of milk is much less likely to cut the teats. Adjust management so that the calf is not fighting for milk.

Advantages of Feeding Fresh Raw Milk

The supreme advantage to keeping cow and calf together is that the calf will be healthy. Of calves that receive no colostrum, virtually none survive. Of those that are bottle-fed with no cow-calf contact, you can expect about 80 to 90 percent survival with good care. Losses are almost entirely due to enteritis and pneumonia. Even when these illnesses do not prove fatal, treatment is expensive and time consuming, and the ultimate growth, thriftiness, and, in the case of a heifer, future milk production are prejudiced by illness in calfhood.

Obviously it sometimes happens, but I have not known a suckled calf to be ill. I don't even recall ever hearing of a bottle-fed calf that was penned where the cow could lick it and touch noses becoming ill. Milk contains macrophages, leukocytes, and an array of other immune factors that are specific to the environment of your own cow and calf. Together with the powerful life-supporting nature of milk itself, this is a formula for success.

Two factors are absent in the life of a separated calf even if you take it warm milk straightaway from the cow. One is the happiness factor. As with all babies, the calf needs lots of interaction to thrive. If there is somebody in the family with time to provide it, well and good. But the mother cow showers her calf with enthusiastic attention and teaches it a lot of cow things. The calf develops a huge will to live.

The second factor is that the cow generates specific antibodies tailored to the occasion if she is in contact with the calf. There is a feedback mechanism dependent on personal contact that enables the cow to produce antibodies to pathogens present in the calf. These antibodies appear in her milk within a few hours after contact. This is like getting a personalized flu shot for whatever strain is going around and getting it before you even know you were exposed. This effect was demonstrated in research more than fifty years ago, but the fact was of no use to the dairy industry, where calves must always be separated, and so it was not widely reported. The same thing occurs in other species.

Because of these benefits, weather permitting, I usually allow the calf to follow the cow as soon as it is a few days old. As mentioned, it appears to eat an amount appropriate to its age and does not get a bellyache. All calf manuals emphasize the importance of early feeding of solids to encourage proper rumen development. Some suggest that an appropriate calf feed be pushed into the calf's mouth to get it used to eating solid food. Maybe there are circumstances where this is worth doing. Or if desired, you could try one of the feed dispensers resembling a bottle, which is designed to encourage the consumption of dry feed. If you do choose to offer grain, avoid corn and soy. The young calf does not produce the digestive enzyme amylase, which is needed to digest corn. Soy contains estrogen analogs that invite a constellation of undesirable outcomes.

A bottle-fed calf, untrained to eat grass, hay, or grain, may go weeks before it ventures on its own to eat solid food or even drink water. A calf at its mother's side imitates everything she does. I have seen a calf trying to eat hay or grass when only a few hours old with no advice from anything but its mother's example. Milk replacers and calf feeds are expensive because they contain high-quality ingredients. Many contain prophylactic antibiotics. You may prefer to avoid antibiotics, and most veterinarians are reluctant to use them without a very good reason. However, a bottle-fed calf is susceptible to scours and pneumonia. Having the calf follow the cow can yield a considerable saving in feed costs and veterinary bills.

Allowing the calf to follow the cow may not make sense unless you have several acres of pasture with plenty of shade and water. Otherwise the calf will get too hot, too cold, and too dirty. On a spacious acreage, be alert for the following problems. The cow, like a deer, knows how to tell her calf to lie silently until she comes back from grazing. If you want to get the calf into the barn for the night, this can be a problem. Only the nudge of the cow's nose will move it. It can hide in plain sight so well that you often don't see it until you literally trip over it. You will need to think about water hazards where a young calf could fall in and drown. Dogs or other predators may attack a calf. And fencing adequate for a cow will often not hold a calf. A calf is apt to lie down next to the fence, then find itself on the wrong side as it gets to its feet.

Management Problems

The management problems associated with keeping cow and calf together include the following:

- The cow, at least at first, may kick during milking because you are not her calf.
- She will bellow when she doesn't know where her calf is and refuse to go out and graze.
- She may try to jump or break down fences to get to her calf and could damage her udder.
- If you have woods, the cow will probably take her calf into them and refuse to come out.
- If you decide to solve all problems by selling the calf after the cow has become attached, she will bellow really loudly for a day and a half, including all night.
- You may have to fight for your share of the milk, or be quite clever, tactful, or gifted at subterfuge to get it.

It is encouraging to remember that for the first ten thousand years or more since the domestication of the cow, people did manage to sort this out successfully. Partial separation with bucket feeding of the calf's share of fresh milk has long been common. Also, the practice of putting several calves on a nurse cow kept for this purpose has a long history and is still done by many. Artificial feeding of calves with total separation from their mothers became a practical possibility only with the advent of modern veterinary practice and the introduction of antibiotics. Without these, the death rate of calves eclipses the advantages of separation.

In general, the cow owner's management task is made easier by a basic attribute of the cow: she is extremely easy to train. Cows really want to do the same thing every day. Once you decide on your plan, after only two or three days your cow will cooperate. The cow belongs to a hierarchical species. Just be sure she knows you are boss, and be quietly consistent. A nudge with your thumb is something she understands, presumably because it resembles the jab of a horn, the method of choice by the boss cow from time immemorial. Nonetheless, the maternal instinct is extremely powerful when given the opportunity to develop and will override just about anything.

Now here is Ann B.'s training method for getting cooperation with letdown. I have used it successfully.

> *Fancy sharemilks quite well — with a foster calf. She holds up for her own calf. I worked out a "reward program" for her with these last two heifer calves (her own and a foster).*
>
> *I left the calf with her for three days, milking twice a day, putting a gallon of first colostrum in the freezer, and then only taking enough to relieve the pressure in the udder. (Note: Partial milking helps forestall milk fever.) On the morning of the fourth day, the calf was separated. After that the cow was milked twice a day, and if she didn't give a full letdown, the calf got a bottle and stayed in her little pen (where mama could reach through, but the calf couldn't nurse). If she gave a full letdown, she got the calf. Only took a week, and she got it through her head that if she wanted the calf, she had to give me the milk first.*
>
> *After six to eight weeks, I gave her a foster calf and went to separating at night, milking in the morning, and leaving the calves with her all day.*

Ann brought her foster calf in together with Fancy's own calf, thus enabling him to learn to nurse. Fancy could not kick him away without kicking off her own calf, so the foster calf got his share.

Training to Lead, Stand, and Back

If you will be keeping the calf, teach it to be led. Put a collar or little halter on your calf. Whereas it is unquestionably true that the calf wearing a collar or halter is at risk of accidental choking, it is also true that without a collar or halter, by about the end of the third day a healthy calf will slip through your fingers and may be impossible to catch. Weigh your personal pros and cons on this one.

The easiest time to train a calf to lead is as soon as it becomes vigorous. This will be by about the third day. Put a lead rope or hay string on its collar to guide its direction, and nudge it on its rump to keep it moving. If it gets confused it will go limp and throw itself down. Rub its back and help it get its legs back under itself. A couple of minutes of this is plenty for the first few times. A line run from the collar around its rump and back through the collar is a useful teaching aid. Then a tug on the lead rope will exert pressure on the calf's rear end and keep it moving. If you bring the calf in at milking time it will want to follow its mother anyway, which makes it easier. Create a little tie-up for it and say, "Stand," while you snap the rope in place. I always teach every calf of either sex to understand the command "Back." This is an important safety measure later on, when the animal has grown large.

Repeat these lessons as frequently as possible, always using the same words. This early training is never forgotten.

While your calf is tied up, accustom her to being stroked all over and having her feet picked up.

Introducing the Heifer Calf to the Milking Routine

Getting your heifer calf accustomed to the milking routine right from the start will pay great dividends later. You can do this whether the calf is bottle-fed or spends its day with the cow, or even if it is a foster calf. Fasten a lead rope and clip near where you milk. Bring the calf along at every milking, tie her in place, and give her a little treat or a bit of brushing. She will soon walk straight to the same spot every time and stand waiting. The little heifer will learn from the start to accept the routine of milking in its entirety. This will serve to ground her mind in the life of a cow. When my cow Fern was milked for the first time, she accepted every aspect of milking as though she had been doing it all her life.

Staking Out the Calf

If it is not spending all its time with the cow, it is easy to stake a calf out on the lawn where it can make good use of your grass. A ground screw with a ring and a rope with a clip at each end is all you need. You can find such things in pet stores; they're made for large dogs. Just be sure to use a swivel clip at the collar connection to forestall choking. Safer, but more work to move, is a stake over which you drop a ring for the rope attachment. For the calf, a halter costs more but is safer than a collar because there is less danger of choking. A halter also offers superior control in leading.

Because of the choking hazard, there is a sound argument to be made for not turning an animal out wearing a collar or halter. On the other hand, there are some good reasons for an animal to wear a collar or halter. For instance, I like my cow to wear a bell. And an animal with a collar is always easier to catch; the collar seems to have a civilizing effect.

An important reason for training the calf to be easily handled is that as it grows you will need to let out the collar. You would be surprised how easy it is to forget to loosen the collar until suddenly you notice the animal looks about to choke. A tight collar invariably makes an animal head-shy and irritable. You don't want a calf roping contest every time it outgrows its collar.

Although a proper cow halter is in many ways preferable to a collar, it is just as susceptible to being outgrown. In either case, leather is a much better choice than nylon; nylon collars and halters tend to cut and rub.

The Calf Pen or Hutch

It is a great convenience to have a little pen for the calf. Five by seven feet is usually about the right size. Simply penning off an area already in use by the cow may be most convenient. Panels made of 2x4s backed by non-climbable wire (closely spaced welded wire) can form a pen that is easily dismantled later. Or just buy "cattle panels." These are versatile, tough, flexible, free-standing fence segments.

There are a few factors to bear in mind when planning the pen: The pen needs good air circulation but no chronic drafts. Both stale air and drafts are invitations to pneumonia. Good drainage is essential. Six inches of sand on a slightly sloping surface is ideal.

You will want easy access to the pen for removal of soiled bedding and manure. If drainage is adequate, the deep bedding method works fine; you can just keep piling in more bedding to keep the calf clean and dry. But you will still need to be able to get at it eventually for cleaning. If you're using deep bedding on a base of sand, make the sides of the pen at least four and a half feet high to accommodate for the rising floor level. And whatever method you use, be sure there are no nail ends or anything else sharp in the pen. I once had a calf tear off her eyelashes on a nail, stripping them right off like they were false eyelashes.

If the calf does not have its mother's companionship, situate the pen where people walk by so that it can receives frequent observation and attention.

Arrange for the calf's feeding containers to hang well inside the pen where the cow cannot reach them.

If the pen is outdoors, the calf must have a hutch. This should face south or southeast in cold weather. Make at least one wall of the accompanying pen of solid boards as protection against wind. In hot weather, the hutch should face whichever direction allows shade and breeze. If the hutch is big enough to be fairly dark inside, the calf can retreat into it from flies. Flies avoid dark places.

A calf needs exercise. It should not spend all its time in a pen.

Vaccinations

Certain vaccinations should be considered. It would be best to confer with your veterinarian as to which are important in your area. The only one I give is against brucellosis for my heifers, which has to be given before six months of age to avoid a false positive if a heifer is later tested for the disease. The chances of a heifer getting the disease are negligible, but if you should want to sell her it's good to be able to say, "Yes, she's vaccinated." She will receive an ear tag with the vaccination, which is useful as a permanent ID number.

If there is a possibility of rabies in your area, cows are susceptible and can be immunized.

The Weak Calf

Some calves are slow to get going. I give vitamins A, D, and E and selenium to any young animals that appear weak. One common cause of weakness is white muscle disease. The treatment is vitamin E and selenium; the two must be given together.

The Sick Calf

No matter how conscientious you are, eventually you'll have a sick calf to treat. The most common illness is scours (diarrhea); see the discussion below. Normal stool is at first yellow and pasty and becomes darker as the days pass. Any diarrhea is a sure sign of scours.

If you suspect a calf is sick, take its temperature. The normal rectal temperature is 100.5°F to 102.5°F. If it is lower than 100°F or higher than 102.5°F, you'll need to start treatment.

"White" or Bacterial Scours

Scours in the very young calf is usually caused by the bacterium *Escherichia coli*, commonly called *E. coli*. Scours are an almost inevitable result of inadequate colostrum in the early hours of life. Colostrum contains antibodies to the bacterial infections the cow has encountered. These antibodies can be absorbed through a calf's intestinal wall only in its first day (or less) of life. Scours cause a whitish, mucousy diarrhea with a nasty pungent smell. The calf becomes dull and listless. Its coat is dull, its eyes sunken. The calf appears to have a bellyache; if it is standing, its back appears hunched. Infection requires prompt action. Typically more than 60 percent of calves with scours die.

If a pharmacy is more convenient than a feed store, get some Kaopectate to help in tightening up the calf's bowels. Similar preparations (mixtures of kaolin and pectin) are available more cheaply from feed stores. With a copious watery stool, the calf will probably be dehydrated, and an electrolyte solution is needed (see below for a homemade version). Test for dehydration by picking up a pinch of skin on the calf's neck or shoulder. It should snap right back when you let go. If it "tents" rather than snapping back, this is a sign of dehydration. Dehydration is the chief killer with scours.

According to a statement by the UC Davis School of Veterinary Medicine, scours in the first few days of life results from colostrum that is inadequate in volume and/or quality; a large volume of *E. coli* that has been ingested, usually a result of dirty teats or bedding; or stress from bad weather or crowding.

It should be noted that *E. coli* is a normal bacteria in the hindgut. Even when ingested, most strains are not problematic if the number ingested is small and other stress factors are not present.

See that the sick calf's bed is clean and dry and that there are no drafts. You'll want to keep it warm, so put hay bales around the calf to contain its warmth, and set up a heat lamp, if you can find a way to do it safely. If the calf is prostrated, consider veterinary help. Glucose and an electrolyte solution are both helpful but may need to be given by drench (tube feeding). If the calf will take its bottle, keep on with its milk but do not mix milk with the electrolyte, as this prevents curd formation, without which the calf cannot digest milk, since it is unlikely yet to have sufficient rennin. Instead, alternate milk and electrolyte feeds. It can be helpful to add rennet to the milk to assist the calf with curd formation. A couple of drops per quart of the liquid is sufficient, or you can get vegetable rennet from a health food store. There is bovine derived rennet and plant derived rennet. Animal derived rennet is somewhat more effective.

It used to be believed that milk should not be fed to a calf with scours, but it is now recognized that withholding food further weakens a weak animal. Probiotics are important. Specially formulated bovine probiotics are available but in a pinch yogurt can be fed.

Antibiotics are of limited value in scours, as most are ineffective against organisms within the gut. However, if the calf suffers from a leaky gut as a result of very severe scours, bacteria may enter the bloodstream and cause septicemia, in which case an antibiotic may be useful.

Electrolyte Solution Made at Home

Electrolyte solution is used for oral rehydration of the calf. You can buy it commercially, but if you have not stocked up or prefer homemade, use the recipe below.

½ teaspoon sodium chloride (table salt)
½ teaspoon potassium chloride (a salt substitute) or 1½ teaspoons cream of tartar
½ teaspoon baking soda
2 tablespoons sugar or molasses
1 liter (1 quart plus 2 tablespoons) water

Combine all the ingredients and mix well. Do not increase the amount of sugar in the solution, as doing so not only will change the osmotic pressure, inhibiting rehydration, but will support undesirable fermentation. To administer the electrolyte solution, it is essential to first warm it to between 100°F and 105°F. A cold liquid will shock the calf's system, forcing the solution into the rumen rather than the abomasum. Give a full quart at least twice a day, separated from milk feeds by two hours.

This formula is suitable for humans.

Other Types of Scours

Salmonella scours are caused by many of the same conditions that cause white scours. Bloody diarrhea is the usual symptom, along with the symptoms listed above for scouring calf. Salmonella is not often seen in the very young calf. Antibiotics are used in treatment, along with the other aids used in white scours. The organism causing salmonella scours can be found in contaminated drinking water and soil.

Parasitic scours are caused by coccidiosis worms or other parasites in the gut, usually picked up by grazing on land that has been heavily and continuously grazed. Calves will develop resistance to coccidiosis, given time. Tablets are available in feed stores that can be put down the throat of a calf to control this type of scours. It may be helpful to feed yogurt or other cultured milk products.

PNEUMONIA

Pneumonia can easily be caused by the same conditions that bring on scours. This is probably why the results of a recent study of calf mortality show that a dry bed is the single most vital factor in calf health after colostrum feeding. But being chilled, particularly on an empty stomach, will also predispose a calf to pneumonia.

At least initially, pneumonia causes a high temperature, reaching as high as 107°F. If not treated, the calf will not last long. The conditions that brought on the disease must be promptly changed. Pneumonia will sweep through a group of young animals if they are confined in close quarters with dampness and poor ventilation.

LUNGWORM

Infection with lungworm (*Dictyocaulus viviparus*) can occur at birth in dirty surroundings or by suckling from dirty teats. The larvae mature in twenty-eight days. The most common symptom is a persistent cough, in England called husk. The infection can be treated with Ivermectin. Garlic in its many forms is used as an organic treatment.

Lungworm larvae can persist for more than a year on pasture. They are found primarily on the lower two inches of grass stems. Therefore avoidance of overgrazed pasture, especially when wet, is a preventive measure. As calves mature they develop an immunity to lungworm.

NURSING CARE

A sick calf must be kept warm and dry. Make it a jacket from an old wool sweater. Set up a heat lamp, if you can do so safely. Prevent drafts. A small sick calf can be kept in the kitchen in a child's plastic wading pool. A bull calf must be set on its feet to pee. And you can make a big difference to a calf's survival simply by keeping it company and petting it. A lonely, discouraged calf usually dies.

Immunity

Another word about immunity: We have already mentioned the special immunity a calf receives from colostrum and the assist provided by extra vitamin A. However, a number of immune factors continue to be present in mature milk and make it an important factor in disease prevention. Milk contains lactoferrin, which binds dietary iron, causing it to be excreted. Iron is necessary for the growth of *E. coli*, salmonella, and staph organisms. By binding iron, lactoferrin prevents the growth of these organisms. Warm raw milk also contains macrophages, immune cells that attack and destroy disease organisms. Furthermore, as long as the calf is actually suckling directly from its mother and in contact with her, another system operates for its protection. The presence of disease organisms in the calf (possibly due to backdraft of saliva into the teat) is detected within the udder. Within five hours, the cow will be producing specific antibodies for the calf and passing them along in her milk. Milk is a biologically active (living) substance.

Daily Monitoring

Perhaps the most useful management tool in raising a calf is just to stop and watch it for ten minutes a day. You will thereby be aware of its state of health,

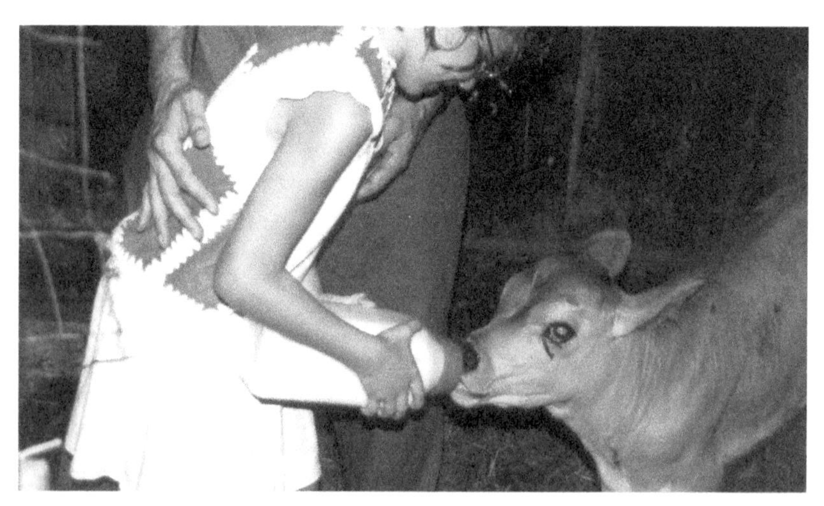

Photograph courtesy of Joann S. Grohman

its growth, and its changing needs. See that the calf always has clean fresh water, a salt lick (the trace mineral type is best – it's usually red in color), and a bit of mixed dairy feed or calf feed containing a balanced mineral ration and extra vitamins. You can take the calf off milk when it is well established on other feed. There are management plans that will get a calf off milk in as little as five weeks, but I do not recommend them. If you are short of milk, just feed milk once a day and make an extra effort to get the calf eating other things. But keep up the milk for at least three months, and preferably longer. The calf weaned too young to inadequate rations soon gets a potbelly and a rough coat.

When a calf stands up from resting, it often gives a big stretch, lowering its back into a U shape. This is a sign that it is feeling fine.

Grain Feeding—or Not

As the calf grows, don't overfeed concentrates (grain). A calf is so cute and appealing there may be a tendency to baby it with feed. It must develop a big rumen capacity, though, and this comes from eating plenty of high-quality hay or grass. Milk does not compete with rumen development. But studies show that heifers that were fed excessive amounts of concentrates produce less milk due to inferior rumen development. Two to four pounds a day of concentrates is the range after weaning. As the calf gets larger, feed more hay but not more grain until it reaches breeding age.

What about total abstinence from grain? This is easily possible if the calf is running with its mother and both have access to quality grazing. Close attention must be paid to the condition of both cow and calf. Merely asserting that grazing is nature's plan does not always translate to sturdy growth. I have seen cows and calves pay a heavy price for idealism in the matter of feeding. That said, there are some sound reasons for skipping grain. These include avoidance of soy and genetically modified feed. Savings in feed costs are a consideration, although this can be tricky; the calf and its mother have to eat something. Finding other adequate nutrition can be a daily challenge and not necessarily cheap. This is the point at which my own regimen falters. I am able to buy a nonsoy mixture called COB (corn-oats-barley) and another that is oats and barley alone. In the fall there are apples and pumpkins. There is further discussion of alternative feeds in chapter 9, "Feeding Your Cow."

Because of the chronic challenge of meeting caloric and protein needs on grass and hay alone and of finding other feeds, I always end up feeding some grain. I am not persuaded that this is a harmful practice.

Height and Weight

To help you evaluate your calf's progress, here are typical weight, height, and chest girth measurements for Jersey heifers of various ages. If you do not have a weight tape, available at most feed stores for minimal cost, you can get a close approximation of the weight of your heifer by measuring around the chest, just behind the front legs, with a cloth tape measure. If your calf is running with its mother it is likely to exceed these weights.

Breeding Age

The standard breeding age for a Jersey heifer is about fifteen months. But if your heifer doesn't measure sixty inches around the chest or weigh more than six hundred pounds, it is better to wait until she reaches the breeding size as well as the breeding age. I usually wait until a heifer is eighteen months to breed her.

Raising a Heifer

It can be profitable to rear a heifer to breeding age or until she is springing (close to calving) and then sell her. You can get an idea of what such a heifer would sell for from what you paid for your own cow. A heifer is usually raised to fifteen months, bred by artificial insemination, and sold to its new owner in the last two months before calving. Alternatively, you could make arrangements for your heifer to go to her new home soon after she calves,

Table 7.1: Calf Growth Chart

AGE	WEIGHT (pounds)	HEIGHT AT THE WITHERS (inches)	CHEST GIRTH (inches)
Birth	54	25.5	26
3 months	130	31.0	35
6 months	275	36.5	44
1 year	515	43.0	56
15 months	615	44.5	60
2 years	800	48.5	66

in which case you get to keep her calf. By the time you sell a heifer you've raised yourself, she will have eaten at least a ton of hay. If you are in a mild climate she will cost you less to feed because she will spend more time on grass. But she will require milk for three or more months to get a good start and a mixed concentrate feed of some kind (for the extra calories) while on milk and through her first winter. Unless you have unusually good pasture and hay, you should continue at least low-level grain feeding indefinitely.

If you are raising the heifer for yourself, you may find it will cost nearly as much as buying another one ready to calve. But you will have the considerable advantage of knowing her background and how she was reared, and also of having the mother-daughter relationship. A heifer will learn a lot from her mother. You also avoid the risk of bringing disease from another farm.

If the heifer is the result of crossbreeding to a beef bull, she can be raised for beef. For that matter, one of the best beef animals I ever had was a Jersey heifer that I could not get in calf. Sometimes a crossbred heifer is just what people want. The lower milk production is preferred by some, and the incidence of milk fever is reduced.

Note: Heifers are often born with supernumerary (extra) teats. Four is the correct number and others must be removed. While the calf is very young you can snip off the extras using sharp scissors held close to the udder. Or you can call a vet and he or she will do the same thing. Puff an antiseptic powder onto the wound.

The Bull Calf or Unwanted Heifer

Compared to a heifer, a Jersey bull calf starts out weighing a few more pounds on average at birth and by age one should weigh about 125 pounds more. A heifer usually sells for more than a bull calf, especially if it is Jersey or Guernsey. If you want to sell, you will need to explore the local market. A Jersey or Guernsey bull calf goes very cheaply and at times can only be given away.

If you decide to sell the calf, give it a couple of days of colostrum feeding before sending it off and be sure the navel is dried up or the calf doesn't have a chance. If you are able to find a local buyer, so much the better. A cattle dealer may know of someone who wants a calf, but the chances are overwhelming that no matter what story the dealer tells you, he or she will take your bull calf, and probably also any unwanted heifer, to auction. Depending upon the current market, it will go from there to a veal operation, to somebody raising grade beef cattle, or to a slaughterhouse, where it will be turned directly into processed meat or pet food.

There is only one fate in store for the bull calf of any breed; the sole question is when and where. Yet home-reared beef animals have much the best life. If the family knows from the beginning that the calf is being raised for beef, emotional attachments are minimized.

The presence of beta-carotene gives a yellow tint to the fat of the so-called colored cattle (Jersey and Guernsey), which causes them to be rejected in the beef trade. (Other dairy breeds or dairy-beef crosses can be raised with confidence of a fair profit.) Because Jersey and Guernsey beef is not commercially interesting, these bull calves won't be worth any more than the slaughterhouse price. Yet this beef is excellent, and if you have the space and inclination, consider raising the calf for your own table. The calf didn't cost you much (only whatever you would have gotten for it had you sold it), and as the months go by you'll see some good beef shaping up. You might be put off by eating what you raise on your own place, but it is the natural way for things to be. Why should you live off the product of some cattle operation miles away? Better that he should have a good life with you and then give back all you have put into him by supplying your table.

When it is time to butcher your animal, there are at least three ways to approach the job. You can butcher it yourself; you can call someone who will do home butchering at your place on a contract basis; or you can have the animal hauled to a local slaughterhouse, which will also cut and wrap the meat for your freezer. If there is such a local abattoir and you choose this approach, visit and find out exactly what they require concerning the hour of delivery and how long they will allow the carcass to hang in their cooler before cutting. Meat is greatly improved by being aged at 40°F for a week before cutting. You will also wish to make arrangements for the liver, heart, and tongue, as these require immediate processing. The butcher (who is usually also the owner) may be able to offer suggestions as to who will truck your animal. You must also be prepared to tell the butcher exactly how you want the meat cut. You may be given a small sum for the hide if you don't want it. You will not be paid for the organ meats if you don't want them.

Small slaughterhouses or abattoirs rarely have overnight facilities for animals, or if they do, these will only be holding pens with no provision for feed or water. These family-operated slaughterhouses often do excellent work – clean, fast, and efficient – but if I am using one, I arrange to bring my animal at an appointed time to minimize trauma to the animal. To really minimize stress, home killing is the only answer. The abattoir should be willing to take an animal that has been killed and field-dressed (eviscerated) and bled, as this is the way a deer or moose will arrive.

In some areas marvelously outfitted little trucks will come to your place and kill and halve or quarter your animal, remove the offal, and wash everything down in about half an hour. Some will even take away the carcass for hanging and cutting. If custom home slaughter is not available where you live, perhaps there is a hunter in the family who will shoot, skin, and eviscerate your animal. In rural areas it is nearly always possible to find someone willing to help with this, given some advance research. It is a perfectly feasible approach, one that has been done for countless centuries. By watching slaughtering I have learned how to do it and have numerous times done all but the actual shot in the head. It is a big job; a block and tackle or tractor with a front-end loader is needed for hoisting the carcass for skinning and eviscerating. You will also need to borrow or contrive a spreader bar for hanging the carcass up by the hind legs. Without this, not only is it difficult to neatly split the carcass, but it is sure to get dirty. A chain saw or reciprocating saw (like a Sawzall) can be used for splitting it into halves after skinning and gutting. Unless the animal is very small or you have strong helpers and adequate space, you will also need to cut it into fore and hind quarters. You can then haul it to somebody's walk-in cooler to hang or, weather permitting, hang it in your garage or cellar. As a last resort you can cut it immediately.

It is important to have a real meat saw, so beg, borrow, or buy one. A chain saw will not do for anything but splitting the carcass: it is too dangerous for close work and too messy and wasteful. Despite these qualifying remarks, cutting your own meat is perfectly possible. You cannot go wrong, in the sense that all the meat is perfectly useful and tasty whether or not it resembles known market cuts. Jersey beef is somewhat leaner than that of the other dairy breeds, and much leaner than Hereford. You may consider this an advantage. The texture is similar and the flavor excellent.

Raising your own beef is well worth doing. You always save money and you know what your animal ate. Kill at eighteen months to two years depending on your personal convenience and hay supply. In an animal younger than that, the beef flavor has not fully developed. But note that immature beef is excellent in its own right.

I don't wish to leave this topic without a warning about slaughterhouses large or small, and this includes ones that are USDA licensed. I have heard a depressingly large number of accounts of cheating by butchers. Possibilities include swapping out your meat for that of somebody else or skimming off your steaks. Against this common theft you have virtually no recourse.

Castration

Because of the possibility of the young bull breeding your cow, you will probably find it best to castrate the calf early, using either the rubber ring or bloodless emasculator method. Castration will be easiest and least painful if done in the first two months. You can purchase a rubber ring tool (elastrator) or the emasculator for less than the price of a veterinarian's call and do it yourself. Follow the instructions given with the tool. Castrating with the rubber ring device could not be easier. Just be sure both testes are in the scrotum when you do it. I once had to call the vet months later to operate on two telltale bulges under the skin. The rubber band operation does not appear to cause the young calf more than fleeting discomfort.

You do not need to castrate a bull to secure good meat. In fact, up to a certain point the animal will grow faster and make more efficient use of its feed if not castrated. There is, however, besides the problem of unwanted matings, an aggressiveness you may find difficult or at least annoying. You may need to carry a stick when in the field even with a steer. A bull is dangerous.

Dehorning

At one time I did not dehorn calves, but now I do. The arguments in favor of keeping the horns – the classic appearance, the cow's self-defense, and the possibility that horns contribute in some way to the cow's well-being – finally gave way to the serious disadvantages. If a cow has horns she will use them, and in the narrow circumstances of domestic life it will most likely be against other cattle (or even against me, in case I have to do something to which she takes serious exception). It is difficult to sell a cow with horns. They can be removed in adulthood, but the job is shockingly bloody and painful and in summer draws flies. I prefer a vet to disbud the calf at an early age. I find this to be the surest and least traumatic for all of us. The elastrator method (just described for castration) can also be used to dehorn, and agricultural supply stores sell a caustic dehorning paste. Additional dehorning devices are available (see the resources at the end of this book).

Weaning

A few more words on weaning are in order. Switching from bottle to bucket feeding is one thing. Weaning of a great big four-hundred pound (or more) calf is quite another. I have tried every method in common use, and while I don't doubt the veracity of my advisors and will admit the possibility that my management skills may be deficient, the truth is that all methods have failed more often than not. The plastic mustaches fall off the calf. The devices designed to jab the udder and cause the cow to kick off her calf are stoically endured by its long-suffering mother; if you reverse the appliance to jab the calf, it quickly learns to flip the thing out of the way. Separation into separate paddocks sometimes works, yet I have known a calf to go straight for the teat after a four-month separation. The only 100 percent effective weaning ploy I know of is to sell the calf. This said, if you are prepared to navigate the weaning issues, there is no question but that a heifer that gets milk and good pasture until weaned at around five to seven months of age will have the well-developed bone structure and excellent health that sets her up for a long productive life, breeding success, and calving ease. Her rumen development and grazing efficiency will reflect the fact that she is living the life for which she was evolved.

CHAPTER 8

Breeding

If you intend to breed your cow, it's important to be able to detect when she's in heat. Ideally, you'll begin to notice the cycles, and mark them on your calendar, and be able to call for artificial insemination in time for the technician to arrive before the heat has passed.

Detecting Heat

At any time from one week after calving onward it is possible for your cow to come in heat. This early heat is sometimes called the "calf heat" and may or may not introduce a normal estrus cycle and may appear as early as six months of age. Occasionally the next heat will be delayed for a couple of months, especially in a thin, high-producing cow. The estrus cycle in cattle is twenty-one days, with a normal range of eighteen to twenty-three days. The actual heat period is rather short, especially in winter. A period of from ten to eighteen hours is considered the length of time in which the cow can conceive. The signs of oncoming heat will start before this critical breeding period.

The signs of heat are mooing and bellowing; agitation, tail swishing, and prancing around; clear or stringy mucus on the vulva or tail; swelling and reddening of the vulva; sudden decrease in appetite or milk production; mounting of other cattle; and a bad attitude. A cow kept without any bovine companions may attempt to mount her human friends. If she is this excited, you will know she is in heat, so maybe you shouldn't turn your back on her that day.

If you have other cattle of either sex, your cow will jump on them and they on her. After a couple of hours of alternate jumping, the one that stands still while being mounted is in heat. This is true even if the other cow appears to be more excited. This "standing heat" – when a cow does not object to holding up another – is the definitive sign of readiness to breed. When you telephone the inseminator, he may ask about your cow's behavior. If you can tell him she is standing for another animal, he will expect his trip to be worthwhile. If you don't have other animals, you can compensate for this lack of an indicator by being very observant in the months after calving. A dairy cow is usually a quiet animal. When she goes into a bellowing session, look for other heat symptoms. As noted, milk production will sometimes drop suddenly when a cow is in heat. A difficulty with this symptom is that you might not react to it quickly. The heat period is rather short, especially in winter. A period of from ten to eighteen hours is considered the window in which the cow can conceive. The signs of oncoming heat will start before this critical breeding period, so it is important to become adept at detecting signs of heat early such that you can arrange for the technician to arrive during this time.

One heat detection method is to stand behind the cow and put your weight on her back end. If not in heat, she will find this annoying. If she stands still and seems to like it, you have a useful indicator.

When you observe the first heat period after calving, note the date and mark your calendar three weeks ahead as well. That will remind you to be on the alert for the next heat so you can establish a pattern. Ovulation occurs about twelve hours after the end of standing heat. It takes some hours in the cow's reproductive tract for the sperm to develop the capacity to fertilize the egg. Breeding should therefore occur toward the end or just after standing heat. Because the length of the cow's heat may vary and the period of viability of both sperm and egg is also variable, the golden moment is sometimes known only to God and perhaps a bull. Yet artificial insemination works impressively . . . when it works at all.

The period of greatest heat activity is between twelve midnight and six a.m. If she has a bell on, you will hear a great deal of ringing, especially if there is more than one animal. Even alone, a cow is much more active when in heat. The poorest time to make observations is at feeding time. Have one person take responsibility for careful observation (just watching the cow) for at least two ten-minute periods each day, and four if you can manage it. Dairymen often report a dramatic improvement in their herd's settling rate with this measure alone. Have everyone in the family trained to observe and report immediately any of the signs of heat. Mark suspected heats on a calendar along with the next projected date. Some cows seem to have a "silent heat" with no detectable signs. Experts writing on this subject suggest the main problem is that we neglect to observe carefully enough.

If all this fails, there is a hormone shot that you or your vet can give that will bring on ovulation (Lutalyse; it must be strictly avoided by pregnant women). Insemination then takes place three days later.

If there is a show of blood, heat is over. A breeding that occurred thirty hours earlier would have been ideal, so you can mark this on your calendar, too.

Deciding on Calving Interval

The widely accepted ideal is to have the cow back in calf seventy to ninety days after calving. With a gestation period of approximately 285 days for a Jersey, breeding at eighty days after calving would produce the yearly calving often sought after. It isn't always possible to be that exact. Yearly calving makes possible the highest milk yields and is a pattern that gives the cow a reasonable balance of milking and recuperating time in relation to the length of the gestation period. You may have a reason to lengthen a lactation so as to be in the dry period at a time when you wish to be away from home. Or if you have more than one cow, you may wish to spread calving around the year. You can also shorten the period between calvings by a month if your cow is young and healthy. Assuming you plan to breed for annual calving, the ideal situation for breeding is to observe one or two heats in a regular pattern and then breed your cow on the next heat that falls nearest ninety days after calving. If you don't observe any signs of heat until eighty days, go ahead and have her inseminated on that heat. See also the section on infertility in chapter 16, "Diseases and Disorders."

There isn't any law that says you must get your cow bred back. You may just keep on milking her through a second year if you wish. Production will decline but will be boosted up again by spring grass. For this to work, great attention to stripping at each milking is required. See chapter 3, "Milking Your Cow."

Artificial Insemination

To arrange for breeding your cow, contact the artificial insemination service in your area, but don't wait until your cow is in heat to locate the service. There are artificial insemination (AI) technicians within a few miles of most places where there are dairies. Consult a veterinarian or a local dairyman about an AI service. There is widespread sharing of semen from proven bulls among the services. Some larger cattle farms do their own insemination, but they'll have a liquid-nitrogen freezing tank for the storage of semen. Sometimes a dairy will permit semen you have ordered to be stored in their tank and even assist with insemination.

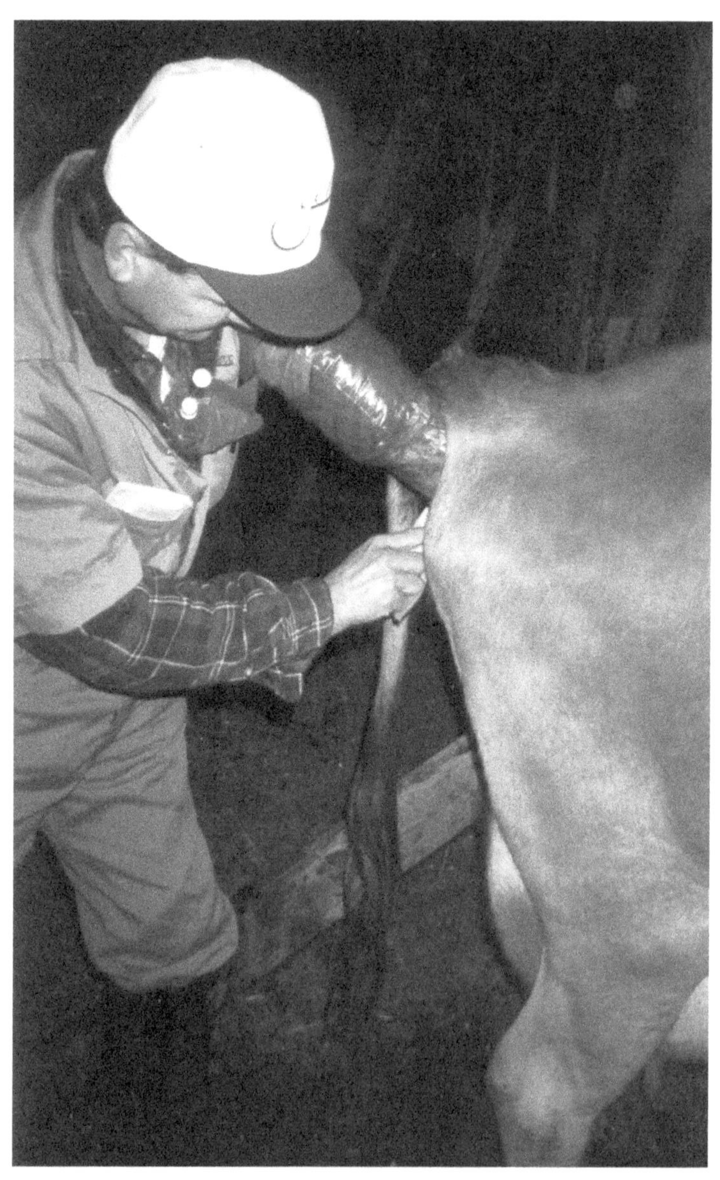

Nathan Cossaboom, Artificial Insemination Technician.
Photograph courtesy of Joann S. Grohman

After you locate an insemination service, it is a good idea to talk to the technician in advance of the time you want your cow bred. Be sure he knows exactly where you live. You can discuss the cost of the service and the bulls he has available. You may ask for literature on the bulls his service offers. The cost of AI varies and generally includes a separate charge for the service and for the semen. There are various arrangements for repeat services if the insemination is not successful. A conception rate of just over 50 percent on the first service is common, so don't be overly concerned if the first attempt is not successful. One of the many advantages of a flesh-and-blood bull over frozen semen is that the bull tries and tries again, and he's there when needed. Furthermore, the AI service semen is greatly diluted. With an inseminator you may need to call by eight a.m. to get service the same day. You may observe your cow in heat at noon and not be able to get an inseminator until late the following day. This could be too late. My experience, however, is that the technicians do the best they can to arrive when needed.

If you want specific characteristics in your calf, such as high butterfat or A2 genetics (see the discussion of genetics on page 125), you can arrange for semen to be pre-ordered and sent to either the AI technician or anyone nearby with a tank. This may add a couple of hundred dollars to the cost of insemination.

Many breeding difficulties (failure to come into heat especially) are reduced when a bull is kept. (The presence of the bull appears to induce heat.) If you have encountered breeding problems in the past and have a bull calf, you may be tempted to rear it to breeding age. I don't advise it. A dairy bull turns dangerous at somewhere around fifteen months of age, and it isn't worth the risk. If you have the space and good fences, a beef bull is safer. The resulting calves will make good beef, and a beef heifer may even be a fair milker. Sometimes there is a bull of a beef breed in the neighborhood where you can take your cow. In point of fact, if he is close enough to hear her bellowing, they are likely to find a way to meet. Then you won't be able to separate them until heat is past. You may miss a milking, always a bad thing, but in this case it can't be helped.

Unless you have been studying breeding records for some time, you won't have much to go on in choosing an AI bull as a mate for your cow. You can tell the technician you want a bull that produces medium-size rather than large calves or one that has a high first-service conception rate or a record of milk or butterfat increase in succeeding generations. But all the technician's bulls will be proven, so if you don't have specific breeding goals, you can safely just ask for the cheapest.

After calling the technician to arrange for insemination, keep your cow in her stall so that no time is lost when he arrives. A big chase is not the best way to get the cow in condition for conception or to make a friend of your technician. A stanchion is an ideal place for insemination. If you don't have one, your cow can be tied in a corner of her box stall. The inseminator puts on a plastic glove and sleeve. He reaches in the anus with one hand, removing the dung as he does so. With the other hand, he inserts a long plastic tube in the vagina, guiding it with the gloved hand (which is on the other side of the vaginal wall) into the cervix. If your cow is properly in heat this will not be difficult. When the end of the tube is in place, the technician pushes a plunger on the stainless-steel holder to eject the semen from a plastic capsule that has been placed inside the end of the long plastic tube.

When he is finished, the technician will write out a breeding receipt, which will include the identification number and name of the bull used. If your cow is registered, he will need to see the registration certificate in order to make out a receipt that will be acceptable in registering a calf resulting from the insemination. The technician will need water to wash, and he will wash his boots in disinfectant. This is a valuable precaution to avoid carrying diseases from one place to another.

It's a good idea to keep your cow in for a while after the inseminator leaves. Evidence indicates that conception is more likely if she remains quiet. Insemination technicians vary in what they tell you about the condition of the reproductive system. Certainly while they have their arm in the cow they have the opportunity to feel the cervix, uterus, and ovaries. Since a technician is not usually a veterinarian, he doesn't offer medical advice, but I have found that an experienced technician can tell me if my cow is well and truly in heat. If she is not, I am onto the wrong twenty-one-day period, and that is useful to know. If the technician should notice something amiss in the reproductive system (such as an enlarged uterus or a cyst), he may suggest veterinary attention.

You might think your cow would object to all this internal probing. She may, but if she is in heat, she will offer little resistance. (Our technician in England, however, was laid up for three months while his arm healed after an irate cow laid it open with a horn.)

A technician is accustomed to serving a cow without aid when necessary, as long as there is an adequate enclosure and the cow requiring his attention is identified (in case there is more than one cow on the place). If you cannot be present for his visit, leave a big note on the barn with your cow's ID number, along with her registration papers if she has them. That is where he will look, and he may not bother to come to the house.

Note the date of the insemination on your calendar and also mark the date three weeks ahead. Watch for signs of heat when this date arrives; if you don't see any, your cow is likely to have conceived. Another sign of successful conception is a sudden drop in milk production after insemination. With good feeding and scrupulous stripping, much of this drop in production can be regained during her pregnancy.

When two twenty-one-day periods following insemination have passed without signs of heat, it is reasonably certain that your cow is in calf. If you want to be sure, wait until eight weeks after service and have a veterinarian do a pregnancy check. A vet does this by reaching into the anus and feeling the size of the uterus. Some authorities say this check can be made after six weeks. Often it can, but it is an annoying waste of a vet call to be told, "I think she's in calf. I'll be able to tell better later." BioTracking (a blood test for pregnancy) is easily possible and can be done earlier if you really need to know. (See the resources at the end of this book.)

If your cow does not come in heat, or if she comes into heat but does not conceive even though inseminated four or more times, read the sections on vaginal discharges and infertility, and consult your veterinarian. If a cow is very thin she may be unable to conceive. Get some fat on her bones.

If there is a special problem in cow keeping that goes with having one cow, it is this matter of heat detection. Sometimes the only practical answer is to forget about calf quality and find a bull. There are always a few cows that consistently fail to "take" with AI but can conceive by natural insemination.

I have known several people who have taken training classes in insemination and learned to do it themselves. This was necessary because they lived in very isolated areas.

A1 and A2 Genetics

For much of the following information I am indebted to *Devil in the Milk* by Keith Woodford (Chelsea Green Publishing, 2010).

It has long been known that casein, an important milk protein, exists in varying forms known as alpha, beta, and kappa. All of these forms are present in the milk of any cow. However, there are a few slightly different forms of beta-casein, with the most important being those known as A1 and A2 beta-casein. In 1993 a researcher in New Zealand, a major dairying country, made the observation that the incidence of type 1 diabetes among Samoan children living in New Zealand was tenfold higher than among children in Samoa. Researchers realized that the major difference between the two populations was the amount of milk in their diet. Subsequently researchers found that the rates of heart disease and diabetes were many times higher in countries where the majority of cows express beta-casein in the A1 form. Active research soon followed, with an emphasis on patenting a reliable test to identify the genes for A1 and A2.

Beta-casein is formed from a chain of 209 amino acids. A1 and A2 differ by only one amino acid. In the A1 sequence, histidine is found in position 67, immediately following isoleucine at position 66; the chemical bond between these two amino acids is weak. In the A2 sequence, proline is found in position 67, and it makes a strong bond with the neighboring isoleucine. The A1 form is believed to be aberrant, having resulted from a mutation around eight thousand years ago.

During digestion, enzymes readily break apart the weak histidine-isoleucine bond in A1 beta-casein, releasing a peptide (protein fragment) called beta-casomorphin-7 (BCM7). BCM7 is an opioid, a member of the morphine family, and is well recognized to have narcotic properties. It is not released during digestion of A2 milk. In susceptible individuals BCM7 is capable of causing a wide range of intolerance symptoms, including bloating, constipation, and nausea. All of these are believed to arise from BCM7 attaching to receptors in the intestines that control the peristaltic movement of food through the intestines.. This is also a well-known effect of some other opioids such as codeine, so it should come as no surprise.

What is even more problematic is that, in the condition known as leaky gut, BCM7 may enter the bloodstream. Leaky gut, more formally known as permeable intestine, is a chronic condition in some people and an intermittent condition in others. The gut of newborn mammals, including humans, is always leaky until a sufficient amount of colostrum has been ingested. The gut of those suffering from Crohn's disease is chronically leaky. During a severe bout of diarrhea anyone's gut may become temporarily permeable. Once in the bloodstream, BCM7 has been shown in rodent studies to attach to opioid receptors, where it causes psychological symptoms similar to autism and schizophrenia. The symptoms are reversed in the presence of naloxone, a well-known morphine antagonist, thus proving the presence of an opioid.

Opioids have been known for a hundred years to affect immune function, possibly explaining the association of BCM7 with numerous autoimmune diseases. BCM7 is an oxidant. It oxidizes LDL cholesterol, a factor in heart disease. Wherever it attaches, BCM7 leads to inflammation, which may explain its association with type 1 diabetes, supposing that it reaches sites in the pancreas. Inflammation is thought to be the initial agent in a great many diseases and disorders and has many causes besides BCM7 from A1 milk. Nonetheless, the evidence against A1 milk is increasingly robust.

Where A2 milk is so labeled, as is the case for some milk in Australia and New Zealand, purchasers often express their gratitude for being able to drink this milk without distress. The issue with A1 milk is entirely distinct from lactose intolerance or classic food allergy. Many people who believe themselves to be lactose intolerant find that they can drink A2 milk. It turns out that the real culprit behind their intolerance of milk may have been not lactose but A1 beta-casein. Alternatively, it may have been due to the BCM7 from the A1 beta-casein slowing down the passage of food in the intestines, and hence giving more time for lactose fermentation.

No genetic engineering or gene splicing would be required to eliminate the A1 variant (or allele) from the national herd. Simple old-fashioned selective breeding would do it. Cows are either homozygous (pure A1 or pure A2) or heterozygous (carrying genes for both forms, in which case their milk is a mixture of A1 and A2). Nearly all Guernsey cows are A2, and many Jerseys are A2; black-and-white cattle are often A1. However, there are too many exceptions for cow color to be of predictive value. All Asian and African cattle that have not interbred with northern European breeds give A2 milk, as do all goats, sheep, yaks, camels, buffalo, and also humans. Three or four generation of using only A2 bulls would virtually eliminate A1 cows. (If you're interested in finding out what type your own cow is, a number of laboratories will test bovine DNA to find out whether it is A1, A2, or a mix of the two. The UC Davis Veterinary Genetics Laboratory in Davis, California, is one such lab; contact them at 530-752-2211 or via their website at http://www.vgl.ucdavis.edu/.)

Milk is a uniquely valuable food. For millions of people milk is the most consistently available high-quality protein food. Although most people, thanks to a healthy digestive system, are probably unaffected by A1 milk, for some it would appear to be a problem of no small significance. Since breeding away from A1 cows is a straightforward matter, why not do it? After initial enthusiasm for pursuit of research on the A1/A2 factor, which has the potential to greatly expand its consumer base, the dairy industry has chosen to dismiss discussion. Keith Woodford informs us that in New

Zealand, which is the world's largest trader in dairy products, herds are quietly being bred away from A1 cows. In Australia, "A2 milk" (i.e., milk in which all the beta-casein is of the A2 form) is now available in all major supermarkets across the nation. A2 milk is also now available in Britain.

Further discussion of these and other issues surrounding A1 and A2 milk may be found in Keith Woodford's book *Devil in the Milk*.

CHAPTER 9

Feeding Your Cow

The basic food of cattle is grass. Sometimes it's that simple. The diet of dairy cows has been the object of at least as much research as has the human diet. Fortunately for cows, the result of all this research is a nutritional level significantly better than what is enjoyed by most humans. But if you are not familiar with bovine nutrition or digestion, a glimpse into any book or even pamphlet on the subject may produce mind-numbing mystery. This has been a factor in making families hesitant to try keeping a cow.

Although goat nutrition is very similar to bovine, myths about goats eating tin cans and Tim Hogan's red shirt have left the impression that goat nutrition is nothing short of entertaining. Goats are altogether less of an economic engine and have inspired less research. Prospective goat keepers have been spared the daunting tables of nutrient analysis. They dive right in.

The cow is a ruminant, as are goats, sheep, deer, elk, buffalo, moose, giraffes, and a number of other grazers and foragers (but not horses). Before food enters your cow's gut, it is fermented in the rumen by beneficial bacteria, which break down cellulose into forms available for digestion. The study of ruminant nutrition has greatly enlarged our understanding of all nutrition and has been of inestimable importance to the dairy industry and to all of agriculture in the United States and around the world. Breeding programs have kept pace, and in fact could be said to have outdistanced ruminant nutrition studies; careful breeding of dairy cows has given them the genetic potential to produce more milk than traditional feeding programs can support. The profit for the commercial dairyman lies in optimizing that extra margin of potential production. You need not worry very much about this.

You Have Choices

Great-grandfather's cow roamed the pasture and ate what was there. At milking time she might have been fed a coffee can or two of whatever grain was on hand along with the apple peelings, corn husks, or watermelon rinds from the kitchen. His cows lived happy, useful lives and produced a creditable amount of milk. They were economic for him, even though his sort of feeding is not of interest to the present-day dairy farmer. I will be discussing something more akin to our great-grandfather's cow management than to commercial practice. The guiding principles are simple and practical and indeed are the same for all dairying.

The choices to make about the feeding of your cow depend on your own time, inclinations, and circumstances. Do you want to minimize your work? There are premixed balanced dairy rations as readily available as dog food from any feed store; just add hay and water. Is it perhaps important to you to know exactly what your cow is eating, so you prefer to work out her ration yourself? It is perfectly possible. You can buy the components of her diet and mix her rations yourself. If you have the land and are able to make the investment in time and at least some equipment, you can grow all of her feed with about the same amount of effort required to grow vegetables for a family; actually, it is easier.

A second conscious choice has to do with the money/milk balance. Depending upon your choice of the foregoing alternatives, this could mean a work/milk choice. The volume of milk your cow will give is greatly affected by what she is fed. You may or may not want to feed her for maximum production. You could feed her five times a day with the highest-quality feeds and get six or more gallons of milk a day. If you do not have any use for this volume, you can reduce the grain and have three or four gallons of milk. Appropriately managed, you can do this without sacrificing the cow's health. You still have to provide enough good-quality hay or grazing so she doesn't lose condition (weight). Dollar savings on grain reductions may be slight and must be weighed against loss of potential profit on her production. To better understand the role of grain in the cow's diet, be sure to read "Practical Cow Feeding" (page 136) and other sections on grain feeding later in this chapter. A cow is less stressed by high production on excellent feed than she is by low production on poor-quality feed. If you are interested in self-sufficiency or intend to make your cow turn a profit, you will be able to use all the milk you can get.

The constituents of the milk will reflect the nature of the cow's diet.

Pushing a Cow

What is meant by pushing a cow for maximum production? Critics often mention this in a roster of complaints against dairying. We need to take a hard look at the concept and find out what it is and what it is not. "Pushing" is not an expression found in dairy publications. Instead we find the concept of "feeding for highest production." One might suppose this is a mere euphemism for "pushing," which sounds like something nice people ought not to do to a cow. Here's how it really works.

Dairy cows have been kept and bred since Neolithic times by people intent on having a docile animal that would produce worthwhile quantities of milk. For much of this history and especially during the last three centuries, high-producing cows have been bred to the sons of high-producing cows. Breeding has been progressively more carefully controlled for milk production, with entirely separate breeds developed to produce meat. With the advent of artificial insemination following World War II, selective breeding took a leap forward with truly impressive gains in productivity. With embryo transplants now commonplace, cow as well as bull genetics have entered the field of play. Today, with less than half the number of cows in the national herd than in 1940, the national herd is producing 28 percent more milk *in total volume.*

Other familiar animals have also had the benefit of selective breeding. Horses have been bred to create draft horses and race horses. Dogs have been bred to pull sleds, course deer, guard barges, and sit in ladies' laps. It is easy to understand that making a courser like a saluki pull a sled would be inappropriate, as would making a fine thoroughbred pull logs out of the woods. At the same time, if you have ever owned a husky, you will know that preventing him from pulling is very stressful for him – for both of you, actually, because he desperately wants to pull. A thoroughbred spending his life in a small paddock with no chance to run is also unhappy. These and other purpose-bred animals are not stressed by the opportunity to do the thing for which they are bred; they want to do it and can hardly refrain from it. The analogy holds up with the dairy cow, because once she has her calf, she can't be prevented from giving a lot of milk. If she is milked, either by her owner or by her calf, and if that milk production is not supported by an adequate, appropriate diet, she will be greatly stressed. For at least several months after calving she will continue to produce a lot of milk, and if not well fed, she will soon become emaciated and may even die.

Returning to the horse and dog analogy, a horse under a determined rider, say Paul Revere, can be made to keep running until it drops dead. Revere did not push his that far, but others have been known to do so. Husky dogs have sometimes met a similar fate in the traces. Most horse and dog racers want consistent top performance, and they achieve this by consideration for the animal combined with a well-designed feeding program. This is the only thing that makes any sense, assuming you are going to keep animals at all. You expect the animal to perform according to its breeding. To compete in the Kentucky Derby or the Iditarod, besides owning exceptionally gifted animals, requires extremely scientific feeding beyond what most people care to get involved in. This does not necessarily imply that either the racing or the feeding is harmful to the animal. However, a performance level satisfying to both animals and their owners can be supported by a good all-purpose diet, and this is all most people require.

Back to cows. To underfeed a dairy cow is extremely stressful to her and is never done in any commercial dairy (although I have seen shocking instances of abuse by individuals with three or four cows). While she may not stay in the commercial herd as long as we might expect to keep our family cow, a cow is not a throw-away animal; she is extremely valuable and is usually treated as such. To get the extra pounds of milk of which her breeding makes her capable requires a computer-designed feeding program. In many commercial dairies this has also come to mean augmenting the feed with components unappealing to human sensibilities and unnatural to ruminants. On top of this, as is not unknown in horse racing, it has also come to include injecting substances to compel performance beyond what diet alone can accomplish. In the case of dairy cows, growth hormones (rBGH/rBST) are used, and unlike horse racing, where this sort of thing is illegal, the practice is FDA approved. *This* constitutes pushing, in my opinion.

You can feed your cow a simple, well-planned diet that meets the needs of good production without pushing or stressing her.

The Rumen

The rumen is an astonishing organ. Once you understand what it does, cow feeding makes sense forever.

The rumen is a fermentation vat, full of microorganisms: bacteria and protozoa. Everything a full-grown cow eats goes straight into the rumen.

The rumen bacteria can do something no animals can do: they can digest cellulose, also known as fiber.

Plants grow by sequestering energy from the sun by the process known as photosynthesis. They add nitrogen, carbon, and water from the air and soil and store the energy as carbohydrates, oils, and cellulose. Cellulose is to plants what bones are to us. It holds the plants upright. The stiffer or more stringy the plants are, such as trees and grasses, the more cellulose they contain. Watery plants like lettuce contain only small amounts. But cellulose is present in all plants.

Cellulose is a major component of the cell walls in plants. Many mammals are able to digest a plant's carbohydrates (starches and sugars) after first breaking the cell walls by chewing (or by cooking), but none can digest fiber. Fiber is not a significant nutrient in humans, although minor fermentation of fiber occurs in the colon, resulting in the breakdown of 5 to 7 percent. In humans and other species lacking a specific organ for the fermentation of cellulose, it serves to create indigestible bulk and attracts to itself digestive by-products, some of which might otherwise be toxic.

Fruits, seeds, and some roots have a lot of carbohydrates relative to the amount of fiber, so many mammals and birds consider them well worth eating and the fiber goes right on through. For instance, we eat apples and carrots and get quite a lot of good out of them. Grass and wood are just the opposite; there is a great deal of fiber relative to the amount of carbohydrates. From the human standpoint, the ratio is so bad that for us to obtain the carbohydrate in grass requires the expenditure of more energy in chewing and passing it through the gut than can be extracted from the grass, plus the mass of fiber causes actual abdominal distress. The result is a net loss of energy; humans starve on a grass diet, and so do all other animals unless they have a specialized organ containing a soup of cellulose-eating bacteria.

Nothing but these microorganisms can digest fiber, the building block of plants. If you see an animal that appears to be living entirely off grass or any coarse or woody vegetation, be it an insect or an elephant, you may be sure it is host to cellulose-splitting bacteria that live in its rumen or other specialized structure.

A major portion of the earth's surface is covered with grass, and so it is not surprising that many animals have evolved to exploit this resource. Grass is a very special plant. Other plants have a central stem off which leaves grow; if their leafy heads are bitten off they have a limited ability to

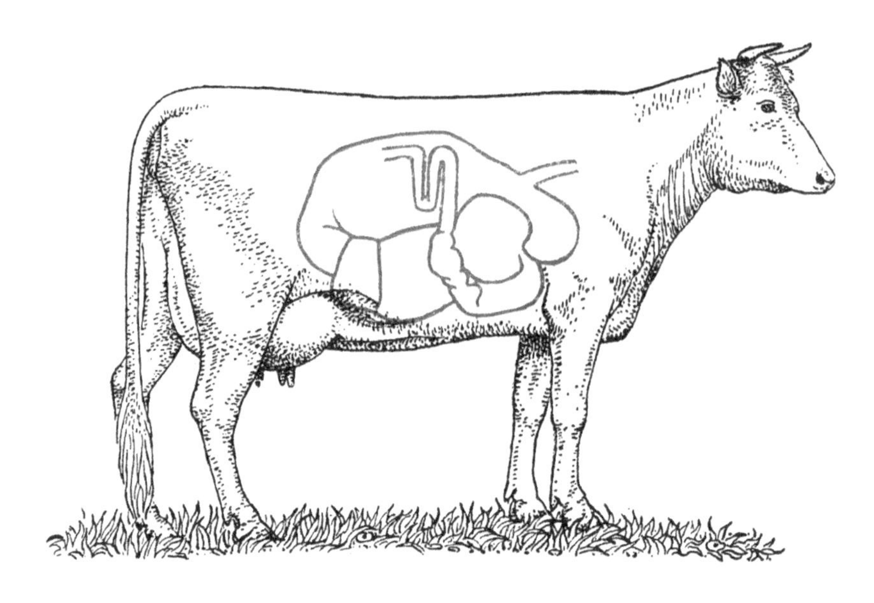

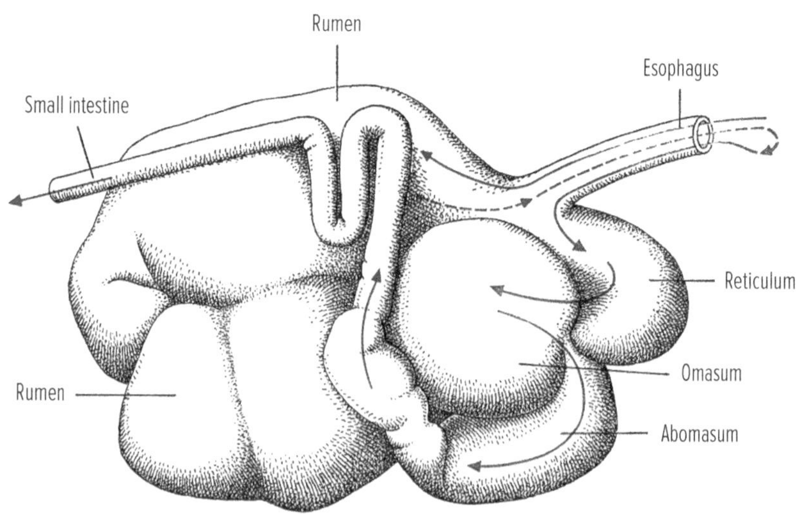

regenerate. But grass grows endlessly up from the bottom, so when its top is eaten off, more comes along. Cattle did not appear on earth until grass was well established.

The typical ruminant is an animal that survives by fleeing from danger. It is a prey species. It goes out in an open, grassy area to graze and eats as rapidly as possible, swallowing its food without chewing. Then it goes off to a safe place and brings up the undigested grass one mouthful, or cud, at a time and chews it thoroughly before swallowing it again. The cud is composed of fibrous stringy material that floats in the rumen like a mattress. Chewing adds saliva and breaks cell walls, releasing carbohydrates, which provide a quick meal for the microorganisms, which then attack the cellulose when the cud is swallowed.

The microorganisms are able to chemically break down the fiber and use its constituents as building blocks for amino acids. These bacteria within the rumen create "from scratch" all the essential amino acids; in other words, they create complete animal protein. They assemble their own cell walls and cell contents from these amino acids.

This complete protein, in the form of a soup of living and dead bacteria and protozoa, passes along into the ruminant's additional stomach compartments for further processing and finally into the abomasum or true gut, where it receives digestion and absorption similar to our own. When food reaches the abomasum, the cow ceases to be a vegetarian; her rumen provides her with a high-protein diet. This is not incomplete soybean or peanut or wheat protein; it is complete protein obtainable to the rest of us only by eating eggs, meat, fish, or other living creatures or by drinking milk.

Every animal requires complete (animal) protein from some source in order to maintain a breeding population. It bears repeating: all animals except ruminants and others with a specialized organ full of cellulose-splitting bacteria must obtain their complete protein by eating other animals or drinking milk. Thus, the bacteria inside an insect digests the cellulose, perhaps a fish eats the insect, then a duck eats the fish, somebody eats the duck, and so on up to the ultimate consumer. But the complete protein was derived initially from synthesis by bacteria. The creatures higher up the food chain accumulate complete protein; they all have to eat other animals to get theirs. Then they pass it along when they in turn are eaten.

The cow proceeds efficiently from resident bacteria directly to her complete (animal) protein diet without the intermediaries necessary elsewhere in the food chain. Because of the extremely simple dietary requirements of rumen bacteria (any fibrous material), the extremely high nutritional value of milk, and the negligible net energy loss that the process requires (it can approach zero), the dairy cow is uniquely efficient among animals, and impressive by comparison with any system. She produces an easily obtainable food of the highest possible nutritional quality right on site from coarse plants of no use to anybody else. The only thing lower on the food chain than a cow is bacteria.

Note that rumen bacteria do not digest one type of fiber: lignin. The large amount of lignin in wood makes sawdust of little value in ruminant feeding. Lignin is also found in dead leaves and brown hay. Only soil bacteria are able to digest lignin.

The Great Urban Myth

Misunderstandings about the natural diet of cattle have resulted from astonishing ignorance on the part of a great many writers and journalists who have not troubled to do their own research and lack the practical experience that might alert them to absurdities.. This has given rise to a collection of urban myths about food conversion (efficiency) rates in animals. One finds such statements as "Fifteen pounds of grain are wasted to produce one pound of meat." Feeding grain is an economic choice. Steers will fatten on grass and cows will produce milk on grass. Wherever this grain mythology is extended to include the dairy cow it become especially egregious; the dairy cow always gives better than she gets.

Because all cattle have been under prolonged and almost entirely misguided attack for several decades, and because you as a cow owner will surely be expected to justify your decision, I have included a scientifically based discussion of all facets of dairy husbandry in the relevant sections of this book, addressing, I believe, every such concern.

Practical Cow Feeding

We now proceed to the practical application of basic rumen function. I am indebted to the publication *Feeding Dairy Cows* by the late Dr. Marshall E . McCullough (W. D. Hoard and Sons, 1986) for the following insight about balancing grass and grain feeding. I have paraphrased him here.

> *In nature's design, the cow would calve in Spring and enter upon her period of maximum milk production. Then in the Fall she would enter a period of reduced milk production with an accompanying need to restore body weight and support another pregnancy.*
>
> *Milk production is best supported with a diet high in crude (plant) proteins, digestible fiber, simple sugars, carbohydrates, vitamins and minerals. Lush Spring and summer growth admirably supports this requirement.*

Six months later in Autumn when plants are in their own reproductive phase, the cow has a heightened need for the extra calories provided by the starches found in grasses and the proteins (nitrogen) and oils of seeds and legumes. Thus the needs of her productive cycle are matched to the rhythm of nature. "You can take the cow out of Nature but you can't take Nature out of the cow."

When the rumen bacteria break down (ferment) fiber, they obtain an energy source for their own use and produce acetic acid, a short-chain fatty acid. Acetic acid is absorbed directly through the rumen wall and is the cow's principal source of energy for all immediate needs, including *milk* and *milk fat* production. *Milk* and *milk fat* are in italics because for most of us, accustomed as we are to thinking of nutritious grains as the seemingly obvious precursors of milk protein and butterfat, it takes a while to adjust our thinking and accept rumen magic; it's grass and hay that end up as milk and cream.

During the writing of the above paragraph I heard an in-depth report on National Public Radio in which it was solemnly explained that the price of ice cream was going up because manufacturers must now pay more for cream . . . cream being more expensive because grain prices have risen, so farmers are feeding less grain and the cows no longer give as much cream. Thus are myths perpetuated. But think about it: wild ruminants give creamy milk, much creamier than what is produced by big Holsteins. Yet wild ruminants get no grain at all.

The carbohydrates in grain are converted by rumen bacteria into other short-chain fatty acids: propionic acid and some butyric acid. All these short-chain fatty acids can be converted to glucose and are the "currency" for all metabolic work in the cow. Early lactation in your modern dairy cow (the only kind of dairy cow available!) requires more energy (as calories) than hay provides, even if she had time to eat enough of it. Without the extra energy in grain she may lose too much body weight.

The protein found in grain does not reach the cow's gut. It is broken down for use as an energy source by rumen bacteria and as a source of nitrogen, which is converted to ammonia and ultimately used as a building block for amino acids.

In managing the feeding of your cow you will be balancing *grass and hay* (fiber, a.k.a. roughage) feeding for milk production and *grain feeding* to supply concentrated calories to provide energy needed to support that milk production and to meet the requirements for body maintenance and reproduction.

If you do not feed grain you will need to find some other high-calorie feed while your cow adjusts her production downward.

Two Tribes of Specialist Bacteria

The bacterial populations that specialize in digesting fiber are not the same ones that digest starches and plant proteins. The various populations explode and crash based on what, if any, food is in the rumen and on rumen acidity. The acidity of liquids is gauged according to the pH scale. It runs from 0 to 14, with 7 the neutral midpoint. Numbers lower than 7 are assigned to substances in the acid range, and numbers above 7 are alkaline. The optimal pH for those bacteria that specialize in digesting fiber is between 6 and 6.8 (close to neutral). The bacteria specializing in starch (grain being the most common source) are most active at a pH between 4 and 5, considerably more acid. The bacteria themselves are usually capable of adjusting the acidity of the liquid in the rumen to their preferred pH based on the type of food present. Saliva, for example, which is generated in great quantity during cudding, is alkaline and raises the pH to assist the fermentation of fiber. The bacteria that lack their preferred food and for which the pH is wrong then crash. If there is *no food* they all crash; it's a big mistake to let a cow run out of food. It can take a couple of hours for bacterial numbers to recover, during which period the cow is totally wasting her time. If the weather is cold, it means her furnace has gone out and she is too bony and short-haired to keep warm. Results can be disastrous.

Besides the bacterial population in the rumen, there is also a significant population of protozoa (single-celled microorganisms) that depend primarily on starch. They operate best at a pH above 6. They perform an important service in the rumen by engulfing starch particles, thus making the starch temporarily unavailable for fermentation. Starch fermentation occurs at a lower pH and if proceeding freely would, by lowering pH, have the effect of discouraging the chief task of the rumen, which is fermenting fiber/cellulose. In this role the protozoa exert a valuable modulating effect.

The acid/alkaline balance in the rumen determines the precedence with which nutrients are digested by microorganisms. In dairy feeding charts you will see the term "neutral" used rather than "base" or "alkaline," because the ideal range for fermentation of fiber (pH 6 to 6.8) approaches pH 7 (neutral); it is in this range that acetic acid is produced by the population of cellulose specialists. The presence of large amounts of starch in the rumen will lower rumen pH by encouraging those bacteria that produce propionic acid. Starch produces "acid stomach."

For the commercial dairyman, this understanding is critical to the task of enabling the cow to produce to her genetic potential. Her feedstuffs must be chosen with care to balance the requirements of energy for milk production (acetic acid) with energy for maintenance (propionic acid). It could be compared with the importance to an athlete of balancing his or her levels of protein, carbohydrates, fats, and salts in order not to crash halfway through a decathlon. The dairyman has to get this right for the cows every day or, like the athlete, their performance shows it. You may not care if your cow gives milk up to her full genetic potential, and you may not care to bother with the requisite mixing and measuring. But some understanding of it will explain the roles of various feeds, and it will permit you to get the best response from the feeds available. You can't help taking pride in the production of your cow.

Grain Cannot Replace Hay

Perhaps the most practical lesson in the foregoing discussion is that if you have poor hay, you cannot compensate for it by feeding extra grain. All you will do is make the cow fat, and the extra grain will actually depress milk production by fostering the growth of starch-loving bacteria, thus creating a habitat too acidic for the fiber-loving milk supporters (acetic acid producers). Better go find some good hay.

Also, do not feed a cow grain on an empty stomach. Make sure she has access to hay or grazing at all times. It is awfully easy to go out for morning milking, let in your hungry cow to eat her grain while you milk, and put out her hay later. If she hasn't had hay beforehand, *both types* of bacteria will crash. Grain tends to leave the rumen rapidly, often before the grain specialists have had time to recover, so it will be largely wasted. The abomasum or "real" gut, which is pretty much like our own, can do very little with grain not thoroughly chewed and mixed with saliva. Remember, the cow gulps, and grain does not come back in the cud. Any whole, unground, unchewed grain ends up as food for the sparrows.

If it isn't convenient to go out at least half an hour before milking to put out hay, make sure you give the cow plenty the night before so she can start breakfast before you milk. If mornings are very dark, if possible have a light on a timer to get her on her feet and eating, being sure she also has water.

Avoid Sudden Changes in Feed

At all times, avoid sudden changes in feed, both in type and amount. To those who have kept horses, this will not come as a surprise. Cows don't colic, but they get bloat. Bloat can occur if a cow is allowed sudden access to a lush field, especially if the forage is cold and wet and especially if the forage is clover or alfalfa, which is a legume like beans, those notorious gas producers. Apart from bloat, which is actually not common except on a cold spring morning, the reason to avoid sudden changes in feed is to give the cow's rumen microorganisms time to adjust. When changing grain mixtures, merge in the new feed by replacing half a scoop a day of the old with the new, building to complete replacement. Otherwise she will not obtain full value from the new feed.

Feeding Hay

The first step in feeding hay is to recognize good-quality hay. Good hay smells good and it looks good. It is leafy rather than stemmy. When you open the bale, the separate segments of the bale, often called flakes or leaves, can be readily loosened up into fluffy hay without raising a cloud of dust. When you offer it to your cow she begins right away to eat it. If the hay is alfalfa it will have stems (alfalfa is a legume, not a grass), but there should be plenty of leaves and they shouldn't all fall off the moment you pick up a handful. If alfalfa totally shatters, then it was cut when too mature and its nutrient value has declined. This is not harmful, merely wasteful.

Commercial dairy farmers usually have hay tested before buying it so they know the relative amounts of fiber, starch, and protein. Some are experts akin to wine tasters; they are able to make an excellent judgment on the basis of appearance and smell and even enter competitions to do so at state fairs. It does take experience to know good hay, just as it takes experience as a shopper to select the best produce and to be able to make a safe guess on which are the tastiest apples or ripest watermelons. When buying hay, shop around for quality as well as price. Open up at least one bale and inspect it for the above traits. Also look for the presence of broad-leafed weeds (forbs). Mature weeds in hay can be a warning sign. The presence of goldenrod, milkweed, and thistle tell you the field has been neglected. Your cow won't eat them if she can help it. She'll try to pick them out but she won't be able to avoid them all. Many are toxic, although typically less so in their dried state, but their presence leads to waste. And the seeds will be introduced to your own pasture in the dung.

On the other hand, many of the broad-leafed weeds are both nutritious and therapeutic. Lamb's-quarter (chenopodium) and amaranth are examples of such useful plants.

If you are confident that you have located good hay, don't hesitate; buy all you can. If it is good hay, local farmers will pounce on it and when you call again it will be gone. If you are uncertain, it may be feasible to buy just a few bales and try it on your cow. Even if she is on good grass, she will usually let you know if she likes the hay. She is an excellent judge of quality.

Unless you have the hay tested, mistakes can certainly happen. Unscrupulous dealers abound. If you get a bad batch, try to locate some better hay and find a buyer for the disappointing purchase. Somebody feeding steers will be a lot less fussy, so cut your losses, and you need not feel you have passed along trouble. A dairy cow deserves the best.

How Much to Buy

I plan to have a forty-pound bale a day for each of the days when there is no grazing. Here in Maine that's six months of the year, or 180 bales. I begin feeding hay in November, even though some grazing is still available, because it is hunting season and Monday through Saturday I keep my cow close to the barn. In late April the fields turn green but I keep feeding hay, because the grass remains too short for efficient grazing and the sod is wet and will be damaged by very much treading.

Timothy, orchard grass hay, or a mix of alfalfa and hay are excellent choices. But fescues and brome are favored in many parts of the country. Each part of the world has its own native grasses and these tend to thrive best.

If hay is the only source of roughage for your cow, expect to feed thirty to forty pounds a day to a Jersey and twice that much to a Holstein. If the hay is of top quality a cow will finish it all, wasting virtually none. If the quality is questionable, as mine often is, I avoid being stingy with it. I let my cow pick it over and use rejected hay as bedding.

If Hay Is of Poor Quality, Cows Eat Less, Not More

Why do cattle eat less of low-digestibility feeds? The reason is that the amount of feed ruminants can eat is determined mainly by the capacity of the digestive tract, particularly the first stomach, or rumen. The cow will eat until the rumen is full. Then hunger signals stop. With highly digestible feeds, the volume in the rumen is rapidly reduced, because these feeds are rapidly broken down. As a result, the rumen empties quickly and the cow eats more to fill it up again. In contrast, a rumen filled with feed of low digestibility empties very slowly, because the feed is slowly digested and leaves large residues (which go ultimately to the dung). Low-digestibility feeds occupy the rumen for a long time, and this limits the amount of feed the cow is able to eat. She may appear satisfied, but milk production and the cow's condition will decline if she is kept full of poor-quality feed.

Knowledge of how this process works will be very useful to you in everything you do in acquiring feed and giving it to your cow. If you make your own hay, you will see the importance of making the best hay you can, given the circumstances of weather, machines, and time. You will need to make your hay early, and quickly, and, if at all possible, in the full sun (see chapter 12 for more on making hay). If you buy your hay, try to find out how it was made. The owner of the hay always knows and may tell you if you ask.

Obviously you do not want just the biggest or heaviest bales of hay; you want the most digestible hay. This is not widely understood. In each of the several places I have lived and kept cows, my family made hay earlier than everyone else. Inevitably a neighbor would drop by, dispensing wisdom by pointing out that if we only had let the grass mature for another week or two, there would have been many more bales of hay from the same land. Certainly true, but we were aiming for quality, not quantity. Once grasses develop seed heads, they send all their energy to the development of the seeds, and the leaves lose nutritive value. We wanted our hay cut before that happened, so that our animals could have the highest-quality nutrition in its most digestible form.

When you buy hay, look for a bright blue-green color (not gray or brown) inside the bale, an aroma of a summer meadow, and the leaves hanging on the stems. If you must feed faded, stemmy hay that lay in the field and got rained on, feed it free choice so that your cow can pick it over, and remove what is uneaten for use as bedding. Avoid feeding moldy hay. If you are forced to deal with moldy hay, feed it outdoors so that the dust drifts away. Moldy hay is less sickening to cows than it is to horses, but it isn't good for them.

I sometimes resort to alfalfa cubes sold by the bag for rabbits. Dried beet pulp is another option, although it is not a substitute for the long fiber of hay. You could also consider making fodder from sprouted grains (see "Fodder Feed" on page 148).

Another factor affecting hay quality is whether it is first or second cut. The first early-summer cutting of hay will be stemmier and somewhat less palatable than that from subsequent cuttings.

How to Feed Hay

You will of course want to feed hay in such a way as to control waste. This is especially important with alfalfa, as even the best of it shatters easily. Remove uneaten hay frequently from the feeder. It can mount up and give the impression she has plenty in front of her when in fact she has already rejected much of it. Leftover hay can be offered to steers or can be used as bedding. If there are a lot of leftovers, try to determine whether you simply overfed your cow or whether she has turned down her hay for some other reason. Never try to impose a "clean your plate" policy on a dairy cow. Try to discern the message behind the uneaten hay. If it is of high quality, she will surely finish it up soon unless she is ill or there is something else wrong; for example, she is out of water. If your hay is from more than one source, stack and feed it separately, noting her response.

As a guiding principle, assume that if your cow eats every bit, you are not feeding enough.

Hay and Grass: The Gold Standard

Hay and grass are a cow's most important feed. No matter what other feeds you use, she must always have either some grazing or a few pounds of hay each day, divided into more than one feed. Because the cow's stomachs are specially adapted for grass/hay, her health will eventually suffer on a diet composed entirely of sugarbeet, pumpkins, or other substitutes, no matter how nutritious. Corn silage, which is composed of the entire stalk including full ears, comes closest to being a substitute for hay, but only if it has fiber length of at least three-eighths inch. Less than that, and cud chewing will not reliably occur and consequently full digestion will not take place. Even if you live in Alaska, where hay is almost prohibitively expensive, you will need at least to give your cow a few handfuls with her other feed. (But if you live in the north, note that alder and willow may adequately induce cud chewing. You will need to observe your cow's response. Asked about the value of alder and willow in cattle feeding, one old-timer in British Columbia said, "Ja, but many die first.")

Grass and hay, with their long fiber, are always the best source of acetic acid, the precursor of milk and the best forage to prompt cudding.

Grain Mixtures

Feed stores sell or can order a premixed dairy ration, which saves much of the guesswork. These feeds are designed to be used as adjuncts to hay feeding. Mixes are available with protein percentages ranging from 12 to 20 percent. Sixteen percent is appropriate for most feeding programs. (This is crude plant-derived protein.) The feed will be available pelleted or in a coarse mix including molasses. It is not easy to be sure what is in these premixed feeds, as the labels refer to the ingredients in generic terms. All the components will have been approved by the FDA for inclusion in dairy feed. The feed mill will be balancing fiber, starch, and protein according to a complex formula. Soy, corn, cottonseed, brewer's grains, beet pulp, and a great deal else can go into the mix according to what is both available and economic for the mill. If any ingredients are from animal sources, this fact may or may not appear on the label, and of course labeling requirements change over time. Dried milk products may be included. Blood meal is permitted in all animal feeds including dairy feed. Bovine bonemeal is disallowed in dairy feed, but bonemeal from swine and horses is permitted. Most premixed feeds include "anti-caking agent," which refers to a type of clay that is added in the hope that it will inactivate mycotoxins such as aflatoxin.

You almost certainly will not be able to beat the price on premixed feeds by buying the components separately, even were you to buy grain directly from a local farmer. Feed prices are clearly subject to the same forces that govern the pricing of a Honduran banana sold in Maine so that it costs less than a local Maine apple. The only advantage to mixing your own feed will be in knowing exactly what is in it. However, mold is arguably the most hazardous feed contaminant, and feed mills are usually scrupulous in avoiding it. Their storage facilities will be superior to any but very carefully designed home storage setups. On the other hand, commercial mixes often include cottonseed meal, a notorious bearer of pesticides.

Virtually all popular brands of premixed feed now include GM (genetically modified) grain. I make an effort to find non-GM feed. Consequently I end up paying more for feed, or else I buy local grains with a lower protein profile.

Premixed feeds are supplemented with vitamins and minerals at levels adequate to forestall most deficiencies. But it is perfectly possible to buy supplements and add them yourself. I always top-dress feed with extra vitamin E in winter and with ground kelp year-round.

Nota bene: I'll say it again – dairy feeds do not and never have contained antibiotics. Antibiotics would suppress those crucial rumen bacteria. Feedlot beef cattle may and do receive antibiotics, hormones, and drugs (such as ractopamine or zilpaterol to suppress fat deposition and compel muscle growth) in their feed or by injection. This is but one of the many ways in which their lives differ from those of dairy cows. In point of fact, dairy cows are often fed *probiotics*. These are friendly bacteria such as those you find listed on your yogurt label.

Hard and fast rules do not (and cannot) exist for the amounts and types of feed a cow must have. Nonetheless, decisions must be made, and the computer has been a vast aid in making these calculations: as soon as fiber goes up, starch may be forced down; there is more than one kind of fiber, and each behaves differently; starches and sugars (carbohydrates), with their potential for causing acid stomach, must be controlled; protein requirements may surpass the amount generated by microorganisms. Fats and oils are a wild card; they are superior calorie sources, yet too much fat depresses bacterial activity in the rumen. Age of the cow, the stage of her lactation, how much she produces, and her stage of pregnancy are all factors that need to be entered into this matrix of calculations. All this nearly defies tabulation. Yet many if not most commercial dairymen do mix their own feed. Computer programs are available to take the mental gymnastics out of it, assisting greatly with these decisions. This could even be fun at home for the dedicated tech user.

I use a simplified feeding program that sacrifices some production: hay and a quality commercial grain mixture in the winter, and grass and less of the same grain in the summer, with vitamin and mineral supplements. I have not found it practical to eliminate grain in the Maine climate. Especially in winter, I feed two to four pounds a day; otherwise milk production falls to disappointing levels. However, there are wide variations among the responses of individual cows.

With some types of management, grain can be dispensed with entirely. If your goal is organic feeding and/or home feed production, your cow's diet will be determined by what you can grow or obtain in your area. If you wish to compound your own feed, you will want to work closely with organic farming organizations in your area to profit by the experience of others. And of course, everybody did it for more than ten thousand years and until quite recently. Quite apart from its other benefits, organically produced milk commands a premium price now even in depressed rural areas, where until recently a dairy operator could find customers only by underselling supermarkets. See also chapter 10, "Your Organic Cow," and the grazing information later in this chapter.

Other Feeds

In additional to the hay and grain that we customarily associate with cattle feeding, an enormous variety of other plant-derived feeds can and are used. If they are ground up and sufficiently palatable, a cow will eat and can digest peanut hulls, orange rinds, pineapple tops and pulp, sugarcane bagasse, tomato and apple pomace, and even, thanks to its cellulose, cardboard and newsprint. Distillers' dried grain (spent mash) is widely fed. All of these and many other items could be included in this nearly endless list and are being fed to dairy cows somewhere. The limiting factors are only that the feed can be broken down in the rumen by bacteria, and that a worthwhile number of nutrients are thereby made available to the cow's gut. Apart from total grazing regimens, it is by using these nontraditional feedstuffs that some dairymen avoid grain feeding. Grain is always the most expensive part of the diet.

No matter what else a cow gets, she must always have long fiber. Twenty pounds of hay or a few hours of good grazing every day are indispensable.

Total Mixed Ration (TMR)

Total mixed ration is a term for an optimal mix of feed, ground finely enough to prevent "sorting" – the cow's just eating what she wants. And as we have seen, a big meal followed by a hungry gap causes great peaks and dips in the bacterial population in the cow's rumen, resulting in inefficient utilization of feed. It won't then come as a surprise that research has shown that a cow will produce significantly better when the exact same ration is fed in six small feeds rather than two big ones. This is an area of management where one-cow production stands to lose or gain a lot.

You will appreciate the importance of the TMR. If everything weren't premixed, it is unlikely that any cow offered free choice would balance it out ideally for herself. Palatability is as great a factor in feeding a cow as it is with feeding a two-year-old child. The persistent myth that, given a smorgasbord, a two-year-old will through the workings of some innate force balance her diet without adult intervention has been long abandoned by most parents. It has also been tried and found wanting in dairy cows. They will eat what has the most molasses in it or tasty sweet corn. And not unlike children I have known, the more assertive cow will get all she wants and may then position herself so that the meek herd member gets hardly anything. (If you put a horse in with a cow he will usually eat what he wants and then spend the rest of the day positioned so as to discourage your cow from eating or drinking.)

The TMR is also the most effective way of preventing the population peaks and crashes of rumen microorganisms. These peaks and crashes waste time and production if, every time separate rations of hay, grain, or other feed are offered, the cow eats on either an empty rumen or one in which the population necessary to that particular feed has declined drastically. Yes, the feed will be eaten. In the case of hay, there will be a lag time for the bacteria to recover because hay will not leave the rumen until it is digested. In the case of grain, much of it may have left the rumen before the necessary bacteria recover, and much of its value will be forfeited. As noted earlier, developing a TMR for a family cow is not out of the question. But most people are satisfied to muddle along much as I do, minimizing the peaks and valleys of rumen activity.

Alfalfa

Alfalfa is not a grass but a legume, and it is higher in protein than all but the best grass. With alfalfa hay it may not be necessary to feed any grain at all. Its long stems also provide fiber. It is high in calcium, and so it is not suitable in the low-calcium milk fever prevention diet, but it is good for a cow freshly calved. It is gas-forming and can cause bloat if it is a cow's only ration, particularly if the cow is grazing it in the field. Give your cow a little grass hay before sending her out on alfalfa so she can't overeat on it.

Alfalfa is high yielding and improves soil by fixing nitrogen, and cows love it. The hay shatters wastefully, so when you're handling it or feeding it, make provision for catching the leafy sweepings. When feeding silage (see page 155) during lactation, a high-calcium feed such as alfalfa should make up part of the ration.

Kelp

During the winter I feed kelp (seaweed). It is available dried and ground. My reasoning is that kelp contains traces of all minerals, so it can make up for any deficiencies in the local feed. Also, iodine is especially important to high-producing cows, and kelp is a good source of it. Kelp can be offered free choice from a permanent feeding station or mixed with grain.

Fodder Feed

Fodder, or sprouted grain, is a relatively new feeding option. Grain is sprouted in trays for about a week, until a dense mat of grass appears, referred to as a "biscuit." The biscuit can be fed to cows as part or even most of their ration. Family cow owners have devised many creative arrangements for producing multiple trays of fodder sufficient to compose much of a cow's diet. Production of fodder requires greenhouse conditions with light, warmth, and regular watering. Growers sprout a variety of seed grains, but barley is a favorite. It is possible to purchase automated sprouting systems capable of supplying a couple of cows or a commercial dairy herd. Cows appear to do very well on fodder, although some hay must still be fed to provide long fiber.

Reasons for choosing fodder feeding include limited access to grazing land, as a response to adverse weather conditions, difficulty in finding or affording organic or non-GM feed, and the desire to provide quality feed at less cost.

Fodder systems are adaptable to confinement feeding. Those who avoid feeding grain are assured that after six days of sprouting, the seed no longer need be considered grain. But it is still not tough enough to qualify as long fiber.

Problems reported with fodder feeding include hesitation on the part of some cows to accept fodder and difficulty in obtaining a consistent product.

Allergens Can Pass through in Milk

Milk from your own family cow seldom induces allergic responses even in people otherwise sensitive to dairy products. But other allergens may come through in milk, as many a nursing mother has learned. Supermarket milk is the blended product of thousands of cows eating an untraceable variety of feeds. Recent studies have shown that peanuts are a significant allergen, for example; and many people already know they are allergic to wheat, soy, or corn. If you arrange your cow's diet to avoid these known allergens, a family member troubled with allergies may have the inestimable benefits of milk and your entire array of home-produced dairy foods.

Common Nutrient Sources in Dairy Feed

It is important to understand the components of the grain you're feeding and the effects they may have on your cow and be able to make informed choices about their sourcing.

Protein Sources

A frustrating thing about feed labels is that they usually identify the grains and proteins generically, without listing their source. The label of any premixed feed should make clear whether any of its protein comes from animal sources, but blood meal may escape mention. Common protein sources include the following:

Plants (20 to 30 percent protein): corn distillers' grains, gluten feed, brewers' grains
Plants (35 to 45 percent protein): soybeans, peanut meal, gluten meal, linseed meal, cottonseed meal
Animals (40 to 50 percent protein): fish meal, meat meal, blood meal, skim milk powder

Mineral Sources

It may also be of use to know the principal mineral supplement sources. For one thing, it makes the reading of labels on feed bags less puzzling. The lists read like a chemist's shelf until you become aware that most of the substances listed are minerals and vitamins.

Calcium: tricalcium phosphate, dicalcium phosphate, monocalcium phosphate, ground limestone, calcium carbonate, oyster and other marine shells
Phosphorus: the calcium phosphates, rock phosphate, defluorinated rock phosphate
Iodine: potassium iodide, sodium iodide, potassium iodate, iodized salt (I include kelp)
Iron: ferric oxide, ferrous sulphate, reduced iron
Cobalt: cobalt sulphate, carbonate
Manganese: manganese oxide
Selenium: sodium selenate

Note: Selenium is usually included in proprietary mixes and with vitamin E supplements, to which it is an essential accessory. It is also present in ground seaweed. Selenium is essential, but it is toxic at high levels.

Vitamins

The vitamins usually included in mixed feeds are A, D_3, riboflavin, B_2, and B_{12} (cobalamin), along with vitamin E and its cofactor, selenium.

Vitamins and minerals are contained in varying amounts in everything your cow eats. In their bulletins, feed companies foster the notion that a cow fed homegrown feed will lack minerals and vitamins. This is not necessarily so. For example, all fresh leafy forage contains vitamin A and vitamin D. Sunlight also directly provides vitamin D. Your cow could be deficient in both these vitamins in winter if she is not exposed to sunlight and if the hay was not properly dried and stored (stored hay progressively loses vitamin A). In northern climates, adding A and D to a cow's diet is a good idea. Most commercial mixed feeds will supplement these vitamins, and you can add them to your own feed by using cod liver oil, which can be purchased in feeding quantities.

Supplementing vitamin E is not necessary when a cow is on grass but is important in winter or whenever she gets only dry feed. Along with its other outcomes, lack of vitamin E leads to off flavor in milk. Wheat and corn germ oils are excellent although expensive sources for vitamin E, but they are not available in feed stores. Read the labels of any feed-store vitamin E supplements and you'll see that they are soy or corn oil with little or no germ included; their vitamin E content usually comes from added dl-alpha-tocopherol, a synthetic form of vitamin E.

Why We Feed Grain

We feed grain for its calories and for the nitrogen in its protein. The calories are in the form of starch and a small amount of sugars. Calories are the single most critical dietary component for every living thing. In the cow the caloric allowance is divided up between her own needs and the requirements of the microorganisms in her rumen. These organisms work hard to produce acetic and propionic acids and the high-grade protein from which they themselves are formed. This work requires energy in the form of calories. The microorganisms take their share first. Therefore we want these acetic acid producers to have enough of an energy (calorie) source so that they do not need to convert plant proteins to energy; we want plant proteins spared for their role as a nitrogen source with which to compound amino acids to meet their own and the cow's true protein requirement (see below). Good hay will contain enough easily digestible fiber and carbohydrates to support microbial activity, with no need for them to bother breaking down plant proteins, which are always a less efficient energy source. When commercial dairymen test their hay and find it lacking easily digestible fiber and carbohydrates, they add a high-calorie food such as beet pulp to the feed so that the plant protein will be spared. A great many dairymen feed a total mixed ration (TMR), which helps sustain the correct population balance of rumen microorganisms and pH. A TMR also prevents "sorting," whereby the cow avoids the dietary components she does not prefer: "No dessert 'til you eat your vegetables."

Because hay is a limited source of free carbohydrates (that's why we can't live on grass), we must also provide enough carbohydrate calories so that the microorganisms that depend on carbohydrates can thrive and produce propionic acid. The cow uses propionic acid as a precursor for glucose. Glucose is the all-purpose energy source for everything. Muscle maintenance, keeping warm in cold weather or dispelling heat in hot weather, switching away flies, chewing, gut digestion, producing dung – all require energy. About two-thirds of all the energy a cow receives is used for this metabolic work. And she still has to make milk, and that too is work. Immediate caloric imperatives will be made up at the expense of other demands. She will lose muscle mass and be otherwise stressed. A cow in peak production has a hard time eating enough to support her caloric requirement and still get enough of that indispensable roughage. Furthermore, the carbohydrates she eats to support all her propionic acid production for metabolic work serve to lower rumen pH below the ideal climate for breaking down the roughage (fiber) for optimal acetic acid production. Dairymen often deal with this dilemma by feeding sodium bicarbonate to raise the pH and overcome acid stomach. You may offer sodium bicarbonate free choice alongside minerals, and a cow will often seem to want it. But recall that the saliva she produces by the gallon during cud chewing is her natural buffering agent.

Meeting Protein Requirements

Grain also provides protein. The assumption used to be that protein was the most important contribution of grain, whereas we now know it is the calories that are crucial (see above). But rumen microorganisms must have nitrogen, and grain is a more concentrated source than grass or hay. Rumen microorganisms must have nitrogen in order to complete their singular task of constructing amino acids. (It is possible to provide free nitrogen from other sources such as urea or ammonium, and this is often done, either to replace grain or to augment nitrogen in other feeds.)

A lactating cow can use about as much protein as she can get. As is true of all nutrients, to speak of "requirements" gives a false view of reality and a false impression that if we just do our mathematics right, natural systems will stand up and say "Yes, sir!" What is clear about protein, however, is that in order for a dairy cow of any improved breed to produce milk up to her genetic potential, more protein must enter digestion in the true gut than any population of microorganisms can generate. This will be true whichever of the popular breeds you choose, because all have had the benefit of selective breeding. With the disappearance of the backyard dairy cow following World War II, you may depend on it: any cow you buy will be the daughter of a good cow, if not a great one, from a commercial dairy that took breeding very seriously.

Note: The "backyard cow" or "scrub cow" advocated by some as possessing unimproved and therefore more desirable genetics is a fiction. If you see a skinny, scrubby cow, she got that way by poor feeding from infancy.

Protein in the True Gut

For her own protein needs, as distinct from those of her rumen bacteria, a cow does not rely on patching together so-called complementary protein from disparate plant sources, each with its own amino acid weakness. Once we move past her specialized stomach compartments and into her true gut, her dietary requirements are essentially identical to those of any mammal including ourselves. To perform her task of producing milk, maintaining herself, and making a calf, only a full array of amino acids as found in animal protein will do. Otherwise the cow will call on muscle as a protein source, like any deprived dieter. If deficits are severe she may also be slow to breed back or may abort her calf, or she may produce a lot less milk than she could, just like any other female. Short of outright metabolic emergencies such as milk fever or ketosis, while other deficits are occurring unseen, what you are most likely to notice first in the case of insufficient protein is that the cow is producing less milk. After the first ten or twelve weeks of lactation, during which she is expressing her non-negotiable hormone-induced impulse to produce, if she is not getting optimal feed including protein, milk production will be forfeited to maintenance.

That is acceptable for most of us one-cow owners. This is what my cow does. I feed her the best diet I can within my limits of cost and availability, and she freshens with five gallons a day, peaks at six gallons, then drops off to three and a half gallons five months into her lactation and looks in fairly good condition. But even with the most scientific feeding it is virtually impossible to prevent a good cow from "milking off her back" in early lactation. She is making up energy deficits using stored fat, and she is supplying protein deficits with her own muscle.

The Cow's Two Protein Requirements

Thus, the cow has two protein requirements to be met. There is the plant protein needed by the microorganisms as their source of nitrogen. And there is the "complete" or animal protein needed by the cow herself. Dairy nutritionists refer to these as rumen degradable crude protein (RDCP) and rumen undegradable crude protein (RUCP). RDCP includes whatever the rumen microorganisms get around to degrading; they are limited in this by the energy available in the feed – energy always is key. RUCP includes whatever the microorganisms didn't get around to degrading along with the contribution made up of microorganisms themselves; they are composed of newly manufactured high-quality protein they have created from the raw material of cellulose and nitrogen. RUCP may also include protein added to the ration with the expectation that it *will* bypass bacterial fermentation and serve protein needs directly via gut digestion. It is therefore often referred to as "bypass protein." This extra RUCP in high-protein feed amendments has increasingly come to mean meat protein. Meat protein admirably serves this purpose since it is what the true gut makes best use of, but rumen bacteria consider it second best. There are billions of tons of otherwise wasted animal protein available from slaughterhouses and other sources, more even than can be gotten rid of in the multibillions of tons of dog and cat food purchased every year by Americans. It makes *economic* sense not to waste it. The practice of feeding meat and poultry by-products to dairy cattle has come under intense fire due to fears about the transmission of disease. The validity of these fears is not fully resolved as of this writing. What is clear is that:

- The American dairy industry as now constituted cannot get along without feeding animals meat protein, because our national cheap food policy dictates that dairy farmers must live by relentless economies of scale. Fewer cows must forever give more milk. As of this writing the only meat product allowed in dairy feed is blood.

- Milk, cheese, and dairy beef (culls, steers, and heifers not needed as herd replacements) compose the greatest source of protein in the American diet and are indispensable. This food source must somehow be maintained.
- Americans do not at present wish to produce milk at home or even encourage it locally, which, were it to occur, would make possible the niche marketing of superior milk at a price spread capable of sustaining dairy farmers while keeping cows at a lower production level . . . slightly more old-fashioned cows.
- When consumers are willing to pay for milk produced without the inclusion of meat or poultry waste as bypass protein, they can have it. My cow doesn't get it and never will.

Fiber

Ongoing research on the nature and importance of fiber in ruminant nutrition has been accompanied by changing terms. The term *crude fiber*, seen in older feeding publications, has ceased to be very meaningful because it fails to distinguish between categories of fiber and embodies an assumption that fiber is to some degree a filler displacing better feeds. The term *roughage* is also used, and the terms are pretty much interchangeable. Studies on the significance of fiber in the ruminant diet have become more coherent and the message is this: fiber is indispensable. The question has become, "What kind?" Ideally, it needs to be the long stringy kind (like grass) and in pieces that aren't too short (those about the size a cow would bite off). Surprising how long it took to work this out.

But wait – that stringy fiber is not all the same. It is composed of varying proportions of pectins, hemicellulose, cellulose, and lignin. These are all cell-wall fibers that enable the plant to hold itself up. Pectins are the most digestible in the cow's rumen, whereas lignin is not digested at all. The proportion of lignin rises as grass matures and becomes drier. So early, lush grass is more digestible than older, stiffer, stemmier grass.

If grass becomes hay and that hay has the misfortune to be made from grass already overmature, and if it then dries too long in the field or even gets rained on, what comes into the barn will have a very high lignin content. Lignin is dark brown, as seen in November leaves. Rumen bacteria cannot digest lignin, so it sits in the rumen for a long time. It depresses the cow's appetite by making her feel full and ends up in the dung. It really is junk food. This is why you can't overcome the problem of bad hay by feeding out more. A cow won't eat while her rumen is full. She may eat more later when her rumen empties, but if no better hay is forthcoming, a cow will lose condition as well as production. If brown, stemmy hay is the only hay available, so long as it is not moldy it is probably better than nothing, but very little should be fed. You may have to buy some high-fiber bagged feed such as dried alfalfa cubes to be sure she gets enough useful fiber. Alfalfa cubes are best presoaked.

What Happened When They Omitted Fiber

As already described, modern breeding for high production has given us a cow whose production potential often exceeds her capacity for food. This has led dairy nutritionists to devise diets unimagined by earlier generations. Also, the logistics of managing several thousand cows make going out to pasture in the old-fashioned way a practical impossibility for huge commercial dairies. Storing, lifting, and serving up enough hay became a nightmare for them. Industrial wastes from the citrus, sugar, and alcoholic beverage industries offer vast savings in money, while huge silos able to serve up silage via a moving belt offer great savings in man-hours. Fiber came to be seen as competing for space in the rumen with more valuable nutrients . . . until research showed a constellation of ill effects from the lack of long fiber. Notable among the ill effects are displaced abomasum, or torsion of this section of the stomach. Usually this problem must be corrected by a veterinarian. Laminitis, a softening of the inner layer of the hoof, may also develop. The rumen wall will lose some of its integrity and become less efficient at absorption. Ulceration eventually occurs due to lowered pH.

Feeding Silage and Haylage

Silage is anaerobically fermented plant material – a process that preserves and enhances nutrient levels – commonly made from cornstalks and other leftovers. Well-preserved silage will have a low pH due to its content of lactic acid. Moist, finely chopped good silage goes down fast and does not prompt the cow to produce enough saliva to buffer the acid. If too finely chopped, silage does not prompt cudding. Cows are often fed bicarbonate of soda with silage to counteract the low pH.

Good silage will have a somewhat pleasing, fermented smell. By contrast, wet, poorly made silage will have a high ammonia content with an odor, and this is as off-putting to the cow as it is to us.

To make good silage you have to make a lot, because there must be tremendous weight to exclude air. The fermentation of silage is an anaerobic process. Traditional silos are tall cylindrical structures within which a great vertical force of material pushes down and evacuates air. Vegetation is finely chopped to reduce the possibility that air pockets will be retained, which would permit mold to grow. The top few feet of silage usually must be discarded due to the formation of mold rather than lactic fermentation. In a trench silo there is even more waste.

Good silage has the advantage that nutrients in the vegetation are well preserved, and actually enhanced by fermentation. And it gets around the age-old problem of trying to make good hay in bad weather. Most silage is made by chopping entire cornstalks, including the ears. Thus it includes both grain and fiber and comes close to being a complete feed.

Hay also may be ensiled and is referred to as haylage; many farmers make it. It permits storage of grass soon after it is cut without going through the haymaking process. It isn't easy to get it right without experience and a silo. A great deal of haylage is now made in plastic bags. The big white bags seal out air and are sometimes referred to as "meadow marshmallows" by farmers. They are wonderful feed and can be delivered in a heavy pickup truck, but the bag must not be damaged. Once opened, the contents must be eaten promptly or will spoil. I don't find them practical for a single cow.

Gene Logsdon, in his book *The Contrary Farmer*, describes a method of making silage in a dustbin with a tight-fitting lid. You pack in grass clippings and add a gallon of molasses. It sounded a lot like making kimchi. It might be fun to try.

Despite the difficulties of making silage, where rain is a constant threat, you should explore the possibility of making or buying silage, especially if you have several animals. Nutrients in good silage are much superior to those in hay that had to dry twice before being baled.

Your extension agent may be able to direct you to a farmer who makes quality silage or haylage. If the farmer doesn't want to sell any, get him or her to teach you the principles. It is one of those skills learned best by participation rather than by reading. Later, after you've seen it done, the pamphlets given you by your extension agent will make more sense.

Good-quality silage smells very slightly sour but not nasty. It should be just moist enough so that a tightly pressed handful holds together for just a moment before collapsing, like bread flour or like snow that's right for skiing. A frequent complaint about silage is that it affects the taste of the milk because of its smell. However, good silage will not have this effect.

Continue to feed some hay along with silage or haylage to ensure sufficient cudding. Feed grain according to your own best judgment as you monitor your cow's condition.

Grazing

What about keeping a cow on pasture alone? After all, aren't cows evolved to eat grass, not hay and grain? The answer is a rather qualified "yes." This is the way most herds are managed in New Zealand, one of the world's foremost dairying countries. A growing number of herds in the United States have changed from confined feeding to mostly grass, based on the New Zealand experience. Here are some of the factors to take into account.

At the beginning of the growing season when grass is very thick, if a cow has all of it she can eat, milk production may be as good as with supplemental feed. Measured over an entire growing season, cows on pasture will not produce as well as those with supplemental feed. The reason is partly due to the fact that grass changes in composition as the season advances, going from high protein/low energy to low protein/high energy and ultimately to stemmy and unpalatable, topped with seed heads that the cow is not designed to harvest.

While this progression by grass in nutrient composition does indeed conform to the requirements of a cow's physiological cycle from spring to fall, the grass tends to go through its cycle a lot faster than the cow unless it is carefully managed. To keep a pasture green and productive there must be enough rainfall or irrigation. It needs also to be mowed to retard its cycle.

Mowing is something a cow does with enthusiasm. But for her to meet her needs on grass, besides being of good quality, the grass must be more than two inches long, preferably seven to eight inches long. At seven inches long she can efficiently wrap her tongue around it and shear it off with her bottom incisors. She does not have top incisors. Picture yourself grabbing a handful of grass and cutting it off an inch or two above the ground with a knife. Shorter grass is much harder to harvest this way. So is long grass after it gets stringy. A cow will keep working at it, but unless her pasture is in good condition (green grass around seven inches in length), she works harder and gets less for her trouble. One result is lower milk production.

If not done by the cow, mowing should be done with a scissors-type mowing machine or a reel mower. A flail-type mower should not be used. It will disperse manure and make a cow very reluctant to graze.

Grass can also be kept down by putting more animals on it. This is very much a judgment call. If, like me, you have more pasture than one cow can eat, she will start off in June with a feast. There is no way that either she or I can keep up with the rate of growth, since I lack the equipment. Soon it is tall and stemmy. But with an appropriate stocking rate and movable fencing I could maintain my cow on pasture alone far longer than I do, and even have untrampled grass growing up for hay. Her production level begins to decline on my regimen.

Dairy graziers accept the lower production level because the inputs are also lower, so the bottom line remains undisturbed, or even improves, and these advantages can apply to you and me. There is less supplementary feed to buy or grow. Most of the manure is left in the field, so there is less labor in barn cleaning and less fertilizer to buy for the fields. There is less equipment to maintain and make payments on, so there is less farm debt to service. Dairy families are thus less vulnerable to vagaries in milk price and so sleep better at night. And veterinary bills are significantly lower; virtually all cow health problems are diminished in a herd on grass.

That is, all health problems except one. A total grazing regimen has sometimes been described as "slow starvation." The cow's condition must be carefully monitored. It's one thing to forfeit some potential milk production. But it is the nature of the mammary gland to keep producing milk as often as milk is removed. That is why you will sometimes see a picture of a nursing mother in a starving human population holding on her lap a fairly satisfactory-looking infant while she herself is emaciated. I have seen pitifully emaciated cows in sparse pastures where they were left to make their own living while feeding a calf; the calf usually looks just fine. Extremes such as this are not likely to occur in a commercial dairy. But the good genetics of the modern dairy cow do prompt her to milk well for many months, and few total grazing regimens can adequately support her, even if she is producing only 60 percent of her potential. If she begins to lose condition, she should be supplemented with grain, her energy food, or with quality hay. (For non-grain choices, see page 140.)

A cow on pasture should never be expected to "clean up her plate," as noted earlier in connection with hay. She needs to be able to make her choice among mouthfuls of grass. Expect her to eat no more than two-thirds of the available grass. This is her nature. You would have to starve her into eating it all. It reminds me of small children. I guess they could be starved into cleaning their plates. Failing that, they seem willing to eat only about two-thirds of what's served to them.

It is not merely milk production that suffers if a cow doesn't get enough to eat. Like any malnourished female, conception rate is impaired. A malnourished cow that is in calf may abort or deliver a feeble calf.

Species of grass vary greatly according to latitude. In the northern United States the native grasses are typically lower in lignin and more digestible. Southern dairymen often plant annuals such as rye grass in the fall to provide grazing. Plots of rye are easily grown on a small scale.

Appropriately managed, a total or near total grazing plan can be the cheapest, the easiest, and the healthiest way to keep a cow. It is indeed the environment for which she is designed. Yet dairymen and family cow owners who choose to pasture feed their cows without offering concentrates must do a balancing act. While it is perfectly true that grass is the natural food of the cow, to produce a volume of milk sufficient to make it worthwhile to own a cow, this has got to be very good grass. Little grass of such quality exists on contemporary farmland except where the farmer has worked for years to improve the pasture. Furthermore, in much of the United States there is no winter pasture. Cows must stand in the barn and eat hay. Few produce much milk on hay alone. There are indeed some parts of the United States, England, Ireland, New Zealand, and elsewhere where grass grows year-round, making year-round grazing possible. Even so, the majority of graziers choose to dry off all their cows during the part of the year when grazing is at its poorest.

The motivation to feed grass alone (in moist temperate climates) or summer grass/winter hay (in cold climates) derives from:

- Saving a lot of money otherwise spent on concentrates
- Saving a lot of time on cow care and giving yourself a three- to four-month vacation from milking
- Optimizing in milk levels of CLA (conjugated linoleic acid) and other long-chain fatty acids and of some vitamins, all of which decline when the cow leaves pasture

For those like myself who prefer somewhat higher production and want milk throughout the year, there is no other choice than to supplement hay feeding with some form of concentrate. Most of the time I feed three or four pounds per day of COB (corn-oats-barley) that my feed store imports from Canada. It is only about 12 percent protein but is non-GM. I supplement this with kelp (½ cup per day) and, in winter, vitamin E.

I urge cow owners who aspire to total grass feeding not to allow attainment of a nutritional ideal to overcome common sense. A cow in a cold climate on hay alone will get very thin. If she is also milking and making a calf she may even die. In summer, her pasture must be sufficiently long and lush to keep her in good flesh. If you want to experiment with total grass feeding, monitor your cow's weight with a weight tape every week. If she cannot maintain her weight, supplement with alfalfa or other high-quality hay, concentrates, or both. I question whether moderate grain feeding (just what's necessary to support weight maintenance) and steady production results in milk of inferior quality.

Estimating Pasture

Here is the rule of thumb for estimating the amount of grazing in a pasture. Place the outside edge of your hand on the ground and measure how high the grass rises against your finger. Measure only the height of the leaves, not the stems; stems have no forage value. If the grass is as high as your first two fingers (one and a half to two inches high), there is approximately 1,400 pounds of forage dry matter per acre. But let it grow. This is too short to graze more than very briefly without damage to the field. It is also too short for a cow to efficiently graze.

When the forage is as high as your fully extended thumb (six to seven inches high), there is approximately 2,600 pounds of forage dry matter per acre. But since one and a half to two inches should be allowed to remain in the pasture, that leaves 1,200 pounds per acre for the cow. This is the perfect height. Much higher, and it will get stemmy. A cow will not select stemmy grass if she can avoid it, so those stems will go to seed, retarding regrowth and lowering productivity of the pasture. This amount of grazing will last a cow about one month if the grass remains in good heart.

Supplementing Grazing

If you need to supplement the grazing, you can try feeding hay, but a cow won't eat it when good grass is available. Ready-mixed dairy feed at 14 or 16 percent protein will not be refused, though. If yours is an all-organic cow, use any organic grain you can get. The main point is to boost energy intake (carbohydrates), and all grains are a good source. Keep in mind the importance of not feeding grain on an empty rumen. If the cow is out all night, she will have grazed before you got up and won't be empty. If she is kept in at night, give her a little hay first thing.

If you're feeding supplemental grain, check the cow's dung for the presence of whole unchewed grain. If they're coming out whole, she's having trouble digesting them. Oats and some other grains can be made more digestible by being rolled or sprouted.

Magnesium Deficiency and Bloat

All cattle are susceptible to magnesium deficiency or imbalance when grazing lush pasture, but calves and young stock more so than older cattle. It causes a paralysis called magnesium tetany, sometimes called grass tetany. It may be accompanied by tremors. If you are in doubt about the mineral levels of your forage, a magnesium supplement is good insurance. Mineral mixtures high in magnesium are available from feed stores or by mail order and can be offered free choice.

Bloat can afflict cattle of any age, with certain animals being more susceptible than others. It is most likely to occur when a cow with an empty rumen is turned out on cold, wet, lush grass. Grazing large amounts of clover or alfalfa can cause bloat at any time. If the night was cold (especially if it was frosty) and the dew is heavy, conditions common in the spring when forage is lush, and the cow spent the night in the barn, then forestall trouble by not letting her out until she has eaten some hay and the frost is off the grass. Hay in her rumen will prevent bloat.

I have not seen bloat occur when the cow spent the night outside nibbling grass at will. Nor have I seen it when grass was three inches high or less, making gobbling impossible.

Bloat is an emergency. Read the section on bloat in chapter 16, "Diseases and Disorders." Study it ahead of time and be prepared.

Extending the Season

Anything you can do to extend the time during which your land feeds your cow will save you money. Kale is a crop able to withstand early freezing. It is easy to grow and produces a lot on a little land. Production is on the order of twenty tons to the acre, although much of this weight is water. Kale is rich in calcium and vitamins A and E, and like other brassica crops it rates high for energy and digestibility. It is a "milk feed," meaning that milk production usually rises when it is fed. One-eighth of an acre could provide fifty pounds of kale a day for three months depending on conditions. This is at least twice as much as a cow should be allowed of kale. Chickens, any other animals you have, people, or your second cow can help eat it. Generally the crop is planted in late June or early July. This is usually after grass has peaked, so that a field could be grazed and then prepared for planting. It can work to plant a field to winter rye in fall, graze it in spring until late June, then put in kale. The crop is ready when your cow comes off her summer grazing, and if the temperatures do not drop below the twenties (Fahrenheit), very often the crop will stand in good condition until the end of the year. Continuous low temperatures will finally destroy it.

Kale and other brassicas (the cabbage family) are highly regarded in England as cattle feed, and they grow well in a cool, moist climate. Except for rape, brassicas are little used as dairy feed in the United States. (Commercial dairymen often feed rape because it is a cheap, available feed in many parts of the country, being the source of canola oil.) All members of the brassica family contain a chemical (glucosinolate) that interferes with the thyroid hormone known as thyroxin. Consequently, heavy or prolonged feeding with these crops can result in a thyroid hormone deficiency. When feeding these crops, be sure to feed some hay, and as a precaution keep iodized salt or free-choice kelp before your cow. Kale and all brassicas are quite high in calcium, so it is desirable also to supplement phosphorus.

Brassicas and other row crops are ill suited to large commercial operations where the logistics of running a herd in and out or cutting and carrying the feed to the cows can be overwhelming. But for one cow or a few, row crops have many advantages. Don't be too daunted by the thyroxin issue. With the exception of grass, the majority of feeds have some disadvantage related to their chemical composition, and even some grasses have toxic phases. Ensuring your cow a varied diet forestalls most problems.

Although it is feasible to cut and carry kale, it is easiest to allow your cow to graze it in strips, using an electric fence that is moved back each day.

Mangels, sugar beets, and turnips also make excellent feed for cows. The tops can be fed fresh, and the mature roots (weighing as much as five pounds each) can be put into frost-free storage. Pumpkins and potatoes can be used similarly. Ten pounds a day of these vegetables is enough to feed a cow along with hay. This will perk up a cow's diet, add a variety of minerals and vitamins, yet not cause any metabolic imbalances. Mangels are hugely productive and highly nutritious for cattle and can serve as a grain substitute.

Any vegetables you feed must be cut into pieces not exceeding about the size of half an apple. They should be fed from the floor rather than from a raised manger. This helps avoid choking, which seldom occurs when food is eaten off the ground. Having no upper incisors, a cow cannot bite into anything big. She needs pieces her tongue can wrap around and place between her molars.

Field corn is a relatively easy crop to grow in most areas. Instead of making silage from it, shuck the ears and allow them to dry in bins made of rodent-proof hardware cloth. Cornstalks (corn stover) can be dried outdoors in shocks (leaned together like tepees and tied at the top). In most climates there will be a lot less waste if these are later stored under cover. Cornstalks can substitute for part of the hay ration. Cows like them, but they are lower in protein and higher in lignin than good hay. Cows do a pretty good job of eating cornstalks even without chopping.

Homegrown feeds including the above and others such as peas and beans (feed cows the entire plant) may be grown much as you might grow and harvest a vegetable garden for family use. Ten pounds of this and ten pounds of that, and pretty soon your cow is well fed. This was the way cows were kept for many centuries, and continue to be in many places, and it works. "Work" is of course the operant word. Your personal motivations for having a cow will guide your decisions on how much, if any, of her feed you decide to grow.

There is one old-fashioned food source for a cow that is hardly any work at all: you can cut down a leafy tree. Alder is a favorite of cows. If there is brushy second growth to be cleared, this can make a lot of sense. Once the trees have been felled, it will allow the cow access to the leaves, and she will clean them up. For this purpose be careful to avoid any species of wild cherry, as it is toxic to livestock.

Water

Water should be available at all times. It is depressing to observe how often a potentially high-producing cow owned by a loving family and generously supplied with expensive feed fails to give the milk she could simply because she hasn't enough water. Providing insufficient water is probably the most common error among family cow owners. A bucket carried by your little boy a couple of times a day is not the answer. Water is serious business for a dairy cow. Plan a system that does not involve any carrying of water or she will be chronically thirsty and you will be out of temper. She needs a great big tub she can walk to every time she feels like it.

A water supply in winter can be a lot of trouble if your barn is cold like mine. I have a tap wrapped with heat tape. It is inside an insulated housing about four feet high. The hose can remain attached and rolled up inside its house.. An electric light bulb stays on inside this box. The entire thing is covered with two horse blankets. Most of the time the system does not freeze.

If possible, warm the water slightly in the winter. There are submersible heaters sold for this purpose (be sure your submersible heater has a proper electrical ground). Too much cold water all at once is hard on a cow. One winter our water supply froze solid for months. The only way the cows could get water was by a long walk to the river, where I chopped open the ice for them. I could manage this only twice a day. Of course, the highest-producing cow was the thirstiest. She was also the thinnest, and several times she went into awful shaking spells when she drank a large quantity. At the time we had no option apart from selling the cattle. But I have read of cows dying from this sort of thing, even in warm weather, if water has been withheld too long. It is called water toxicity.

A cow needs at least ten gallons of water to produce five gallons of milk if she is on dry feed, plus additional water, totaling as much as thirty gallons, to meet her other physical requirements. On green grass she requires somewhat less due to moisture in the forage. Much of the water goes to produce saliva, which she mixes with her cud as she chews. If there are critics to be answered regarding the water consumption of dairy cattle, note that this water does not disappear; aside from creating valuable milk, water consumed by a cow is returned to the land as urine and manure, and it becomes part of the natural cycle of nutrients. It does not become part of the polluted waste stream as occurs in cities. It is not lost.

A fifty-gallon plastic stock water tank from a farm supply store is a good investment. Float valves are available so that the tank can remain dependably full. In warm weather it will need frequent cleaning to get rid of algae. I wouldn't recommend galvanized water tanks. As rain becomes more acid, there is a potential for a reaction with the zinc surface. Zinc in this form is poisonous. The common practice of adding a little vinegar to the cattle tank is inadvisable in a galvanized tank.

At all times I keep a board or branch in my water tank. This way, I never find little drowned animals.

A cow is thirstiest right after milking and will usually head straight for water. If there are several animals that must compete for the same water supply, attend carefully to the possibility that the boss animal is intimidating the others. A cow interrupted in quenching her thirst by reaching the end of the supply, or anything else for that matter, will usually not resume drinking again for a long time, and milk production will suffer. Always treat a cow's drinking as a sacred moment, one in which she is not to be hurried. Try not to allow anything to interfere with her nice long drink. Instead she will go in search of food. After all, *she* does not care how much milk she makes.

A cow will not walk a long way for water unless she is extremely thirsty. She will walk much farther for food. So if she will be walking five to ten minutes away to graze, it becomes especially important that her drink is ready and waiting before she sets out.

Water supplies in most parts of the country are now at risk for nitrates. If you have a private well, you will want to have it tested once a year for nitrates for both your own safety and that of your cow. They may come from runoffs of agricultural applications or from contamination by septic systems.

If you must use chlorinated water, either install a dechlorination system or at least get a water tank with a broad surface area so that the chlorine has a chance to dissipate. Chlorine suppresses rumen bacteria. Dairy farmers also defluorinate their cows' water if fluoride levels exceed 1.5 milligrams per liter.

Farm ponds also need evaluation. If manure can get in the water from animals grazing nearby or from waterfowl, or if the water is stagnant or contains decomposing vegetation, it becomes easy for coliform bacteria (*E. coli*) to exist at a high level. As discussed earlier, the common strains of *E. coli* are not toxic in the colon, where they are meant to live, but if consumed in large quantities they will compete with proper rumen bacteria, making fermentation less efficient and giving a cow a chronic stomachache. Milk production will of course be affected.

An emergency water supply that has been a lifesaver for some people, whether for farm or domestic use, is an old milk tanker truck filled up with water. A milk tanker will have always been kept clean, unlike other tankers, which may have been used for anything.

Cows know when water is clean and free of chemicals or bacteria, and when it is clean they drink a lot more.

Stray Voltage

Stray voltage is a very serious problem around cattle or any four-footed animal that stands on bare earth or damp cement. It is much more common a problem than you would think. Electric companies have been reluctant to address the issue and may even deny the possibility. Take great care that all wiring near cattle is properly insulated and grounded. Electricity can travel a surprising distance through soil. Animals are rarely harmed in any great way by stray voltage, but they may become very reluctant to walk through a muddy spot to the water tank or be hesitant to drink because they are getting a buzz. If the water tank appears crystal clear but your cow won't drink from it, suspect stray voltage. It usually cannot be felt by humans, but a cow's nose is especially sensitive to it.

Salt

A cow needs salt available every day. If she lives in her stall, you can buy a small block and a holder to attach to the wall so she can lick it anytime. Outdoors, the easiest method I know to make salt available is to get a fifty-pound block. The block will have a hole in it. You simply drive a pipe into the ground and then fit the salt block over it, with the pipe in the hole. This will last for months. Rain will erode it to some extent. I always buy the red block with added minerals. A cow won't get enough minerals from the block to meet all her needs.

Methane Production by Ruminants

Methane, because of its role as a greenhouse gas, has received a great deal of attention. Its production by ruminants is widely misunderstood. It is a normal by-product wherever fermentation occurs. I introduced the following discussion in the *Maine Organic Farmer & Gardener* quarterly for June 20, 2010.

> *Methanogens, a life form smaller than bacteria known as archaea, do part of the work in the rumen. Methanogens produce methane (CH_4) when extra hydrogen is left over following less efficient fermentation of cellulose by (other) rumen bacteria. This occurs when cellulose, a form of glucose, is split. Methane is also known as swamp*

gas or natural gas. If you cook or heat with gas you are using methane. Like CO_2 from oil or coal, the methane we cook or heat with lies stored in the earth and remains inert until mined and released. Megatons, trapped following ancient fermentation, are now being released from tundra by melting permafrost. Oceanographers now describe vast belches coming up in the Bering Sea and South China Sea. In some places the sea is foaming like a shaken soda, the methane is emerging so fast. Anthropogenic (human derived) sources of methane are rice paddies and landfills, both of which emit more methane than does livestock. The rumen is a controlled fermentation vat and produces methane at a modulated rate.

Methane contains energy, as you know if you cook with it. The amount a cow's rumen produces varies according to diet. As mentioned above, the rumen is designed to ferment stringy cellulose (grass, hay), not grain, and consequently much of the grain a cow eats is passed unaltered to gut digestion, similar to our own, and and is not fermented. For this reason a diet high in grain results in proportionately less methane release compared to grass or hay, which always must be fermented in the rumen. Forbs (broad-leafed plants) found in natural pasture assist with fermentation and boost its efficiency so that less of methane's energy is lost to the cow. Like the buffalo before them and the deer in the woods, grazing cattle will always belch up excess methane. Ruminants produce more methane than other plant-eating species because their large rumens are actively breaking down more cellulose, much to our benefit. This does not unbalance the universe and never has. It feeds us.

Cattle, whether beef or dairy, if eating diets high in grain, will, as noted, produce proportionately less methane than do cattle on grass or hay because much of the grain fails to be fermented but is instead passed intact to the small intestine for standard carbohydrate digestion. However, the practice of collecting manure as slurry in vast lagoons produces methane by the ton. These lagoons are also the mode of collecting manure from swine and poultry CAFOs (concentrated animal feeding operations); thus swine and poultry, despite being nonruminants, also become responsible for methane production. Methane production is anaerobic (without oxygen). Manure lagoons crust over on top and the slurry does not circulate; these lagoons provide ideal anaerobic conditions. It is these lagoons, not cattle themselves, that are the chief source of methane now being attributed to livestock production.

Upon first consideration it might be supposed that if cows were out grazing on pasture the methane contribution from manure would be the same as in confinement, merely spread out over the countryside. This is not the case. Cow patties dropped in the open air on pasture result in no methane production. Cow patties in the open air do not support methanogens. They are consumed by insects, birds, and aerobic soil bacteria.

Beyond its role as a greenhouse gas, methane remains an important energy source. Methanogens (the bacteria that produce methane) have many important roles. Archaea in the seabed, for example, have been found to play a central role in the planetary nitrogen cycle on which all life depends. Methane is at risk of becoming demonized before it is properly understood.

So why don't we move cows out of feedlots and dairy barns and onto pasture where they can participate in the short-term carbon cycle and carry on belching harmlessly like their ancestors? Quite aside from resistance from agribusiness, which prefers things as they are, many well-respected writers and scientists dismiss this as a practical impossibility. Insufficient land is the usual excuse. Comparisons of food calories per acre between animal and vegetable production are always mentioned. Academic studies consistently state that cows cannot be pastured locally on grass in numbers adequate to meet consumer needs. But until it has been attempted, nobody is qualified to make such a statement. Well-known Maine farmer and gardener Eliot Coleman has repeatedly demonstrated that production from a vegetable garden is by no means finite and can be impressively greater than most people realize; the same is true with cows, no "pushing" required. Estimates of land requirements obtained by dividing the number of people into USDA statistics for farmland acreage make it sound hopeless to depend upon local food production of animals or vegetables. These linear production models, seldom challenged, have formed the basis for assumptions about the potential for meat and milk production by influential environmental writers and researchers, few of whom have a cow in the backyard. But there is no linear relationship here. The upper limit for integrated local food production of plants and animals depends on dedication and imagination and is not known. When free-market forces are allowed to operate, food production soars. We do not currently enjoy free-market conditions. The very fact that so many local growers are already flourishing under stifling constraints hints at what we may look forward to if and when some of the more onerous regulations are eased.

Coleman questions the assertion that animal agriculture has anything to do with global warming. He suspects that oil interests and corn and soy producers, along with vegetarian cheerleaders, are feeding us disinformation about the role of animal production as a factor in climate change. It is in fact plant crops that are responsible for displacing small farms in both the United States and Africa and for deforestation of lands in South and Central America and in South Asia. Cattle are used as a quick cash crop before the land is dragged clear for corn, soybeans, or palm oil. Statistics regarding the contribution of animal agriculture to greenhouse gas production are clearly being manipulated for somebody's benefit.

Here are U.S. EPA figures from 2007; ratios change little from year to year: In 2007 U.S. agriculture accounted for 6 percent of U.S. greenhouse gas emissions. Of this 6 percent, 24 percent was from enteric (rumen) fermentation by cattle (excluding manure management). So 6 percent × 0.24 = 1.44 percent. If all cattle were killed, then 100 − 1.44 = 98.6 percent of U.S. agricultural greenhouse gas emissions are still with us. Absent methane emanating from manure lagoons, the methane contribution of cattle (called enteric) is negligible. The real sources of greenhouse gas emissions from agriculture are manure lagoons and the petrochemical inputs to plant crops. Declarations about environmental damage by cattle are based on our current unsustainable animal husbandry practices involving CAFOs.

CHAPTER 10

Your Organic Cow

Of the many excellent reasons for keeping a cow, quality control of your dairy products heads the list for most people. Following more than a century of advocacy, proponents of the all-organic diet have, I believe, carried their point.

Use of the term *organic* as an adjective to describe plants and animals thriving on composted soils was popularized in the late 1940s by J.I. Rodale Sr., founder of *Prevention* magazine. The term caught on at once with the public, much to the dismay of the scientific community, for whom the word *organic* has a well-established definition in chemistry as "a molecule containing carbon." The USDA now owns the term *organic* and has created a strict legal definition. But as a cultural concept, rearing chemical-free animals and plants has a far longer history than does use of the word *organic* as a nutritional descriptor. The conviction that fresh, whole foods grown on properly nourished soils are the foundation of good health, good physical development, and long life had been already clearly enunciated by the middle of the nineteenth century.

These precepts are now encompassed by the term organic, and most of us non-chemists feel we know what we mean when we say plants or animals are "organically grown." Our definition is little challenged when all it involves is the choice of which prepacked carrots to put in the supermarket cart. It will require some thoughtful consideration when we are feeding a cow.

For many people, organically grown food is defined negatively: the soil did *not* receive chemical fertilizer from a big plastic bag. The crops were *not* treated with chemical pesticides and herbicides. Yet fully eighty years before the word *organic* was first applied to food production, the whole-foods movement was active under a variety of names in the United States, England, and Europe. What all had in common was a positive belief in the importance of soil improvement. To anyone directly involved with organic food production, this emphasis needs no defense.

As enthusiasm for this positive approach to food production reached critical mass among consumers in the 1950s and '60s, it met a great deal of antagonism from farmers, schools of agriculture, and the USDA, all of whom had been congratulating themselves for coming up with chemical methods that would end world hunger once and for all. It met equal hostility from a burgeoning fertilizer and pesticide industry, which for the first time felt itself compelled to respond to criticism and justify its methods. With cooperation from the USDA, Extension Service plots using chemical methods were planted side by side with plots that received no fertilizers or pesticides at all. The untreated plants were miserably small and worm eaten. Organic growers were then derided with the question, "If we don't use chemicals, which third of the world's population shall we allow to starve?"

This doomed research had asked the wrong question. The contest was never between chemistry and some depleted tract alongside. The contest has to be with soil enhanced for several years with compost and/or animal manures, plant materials, and carefully selected amendments such as lime and greensand. Pesticides, if any, are selected from among those that cannot affect people and that biodegrade completely. The overarching point is that properly fed soil results in big, healthy plants that resist insect damage. These plants in turn support big, healthy animals, such as your cow. Furthermore, and not to be ignored, well-cared-for soil gets better every year and generates no environmental debt. Chemical agriculture can achieve, as any trip to the produce section of your supermarket clearly displays, some very beautiful results. But it also creates an environmental debt in depleted soil, contaminated water tables, and illness among farm workers. To my mind, it is not organic advocates who should be required to defend the question, "Which third of the population shall we allow to die?"

Since "organic farming" based merely on avoidance of chemicals doesn't work, and since there are hundreds of ways to improve soil, how do we know when we are doing it right? Some of the soil amendments widely accepted as "organic" are the very same ones often used in orthodox, or chemical, farming. Who decides what is acceptable? The USDA has now established certification standards for organic agriculture, and these have resulted in a reluctant truce among members of the various organic growers' associations. A truce is not the same thing as truth. But it serves to back up marketing and build consumer confidence. Associations in states such as Maine, California, and Oregon have achieved a strength and reputation for integrity that makes their endorsements meaningful. Individual growers in every state make a serious effort to use methods that conform to most organic tenets.

With USDA certification, *organic* now has a legal definition. However, when applied to animal feeds it remains problematic. Poultry or swine, in order to thrive and reproduce, have a dietary requirement for the full complement of essential amino acids, as is found in animal protein. To meet the organic standard these animals would have to receive a diet that included meat or milk protein from other organically produced animals, or fish meal from wild-caught fish. Not only would this make for a prohibitively expensive diet, but because use of recycled animal products in feed is now largely phased out due to concerns over "mad cow disease," most such choices are disallowed. Instead, feed is fortified with isolated lysine and methionine, the amino acids in which plant foods tend to be deficient, thus creating a complete protein synthetically. Such feed and the resulting animals and eggs are then touted as having been produced with an "all-vegetarian diet." This appears to have been a successful marketing ploy but skirts the truth. This use of amino acids is not permitted in food intended for human consumption. Cows, on the other hand, can easily be fed a truly vegetarian diet of plant products because complete protein is produced for them by rumen bacteria.

Organic Is Easier with Cows

Cattle, sheep, and goats are ruminants and generate their own complete protein, as explained above. They need eat nothing but plants, so providing them a diet wherein everything is grown on improved, "organic" soils becomes more feasible. You can do it yourself if you have the space. As I hope the preceding discussion has shown, that may be the only way you can be certain that your feed contains nothing you disapprove of.

Controlling Allergies

If somebody in the family is seriously sensitive to environmental contaminants, keeping an all-organic cow is an incomparable way to start building him or her up, as you build up your soil and improve everything you grow. As one who has lived with allergies in their many permutations, I have long been convinced that whereas eliminating allergens may be a necessary first step, what works best in the long run is to build up the general health of the sufferer. The allergies then diminish or vanish completely. Few people who have been declared allergic to milk have any problem with fresh milk from a family cow. None of the substitute foods offered allergic persons is the nutritional equivalent of milk. After temporary relief on an elimination diet, symptoms usually return and health continues to decline as the victim eliminates yet another wholesome food.

My Own Approach

My way does not meet USDA's national organic standards but could presumably be made to, had I the resources to become certified.

My cow gets hay made on my own fields, which have never since the land was cleared 150 years ago received any pesticides. Because of the decline in farming, this is the case with every hay field and pasture around where I live in Maine. In bygone times, many fields received animal manures. Most now are merely mowed, resulting in a predictable decline in hay quality.

I have spread lime and I have brush-hogged yearly and left the grass behind as mulch, so my fields are somewhat improved. My hay crop remains thin but nutritionally is within the range of good. If weather and equipment permit making hay under optimal conditions, my hay will be better than hay from a healthier field that was cut too late or was rained on.

The most important thing I have done to improve my land is keeping it stocked with cattle and sheep. My pasture has shown steady improvement compared to another field treated identically but without the benefit of livestock.

My fields are divided so my cow does not get hay from a field where she has grazed within the last three years. This helps break the cycle of parasites.

I don't compound my own grain mix, so I can't know exactly what is in the mix I buy. I read the labels, of course, but some information will never appear because the FDA does not require it. As on foods for human consumption, the label will include the statement "from one or more of the following sources . . ." The brand my cow likes best costs a few cents more but has a good reputation, so I buy that. The label does not state that the mix contains any protein from animal sources, which is to be avoided.

I feed COB (corn-oats-barley) much of the year. It is lower in protein than commercial feeds, and its only additive is a little molasses to control dust. If high production is not your aim, COB will help maintain your cow's condition.

Clarinda gets a couple of handfuls a day of dried ground kelp and a free-choice mineral supplement. I pour two glugs and a gurgle of "wheat germ oil" on her grain at every feeding. I put that in quotes because it is hard to find any pure wheat germ oil except in health food stores. What you get at the feed store is soy or other oil fortified with a little wheat germ oil and vitamin E.

Clarinda gets lots of apples in the fall and any other tidbits I can scrounge. See chapter 9, "Feeding Your Cow," for tips on crops you can grow for your cow. If the effort is important to you, you really can grow all your cow's feed and know exactly what goes into her. As you read in chapter 9, you can, with some lowering of production, keep a cow on pasture and hay alone, and properly managed she need not suffer nutritionally. It's all in the quality of the soil the crops come off, and *that* is the part of putting pure organic food on your table that takes work and keeps the mortgage on your environmental future paid up.

The feed choices in chapter 9 are equally valid for your all-organic cow, provided you can locate organic sources. At present, obtaining organic feed is keeping a ceiling on expansion of organic dairying, despite pent-up demand.

My commitment to the organic concept goes back more than four decades. But if I must choose between ordinary feed or no feed, I don't let my cow go hungry, the same principle I apply to feeding the family. Avoiding otherwise appropriate feeds because they aren't organic can become an intellectual trap.

Medications

USDA Organic standards do not allow antibiotics for dairy cows. Organic dairymen must grapple with the problem of what to do in case of mastitis or other illness. Many have found homeopathic veterinarians, but few I have interviewed find this to be a wholly satisfactory answer. I don't have personal experience with homeopathy. I do have considerable personal experience with the following health problems; in fact they are the only health problems that have presented themselves to me in keeping a family cow. We saw a greater variety of illnesses when we had a sixty-cow dairy.

MASTITIS

Eventually almost every cow develops at least one mastitic quarter. Read the discussion of mastitis in chapter 16, "Diseases and Disorders," and be alert for symptoms so you can act quickly before the infection gets far advanced. The important thing to do is to keep stripping out the milk. The cow may kick; mastitis hurts like a bad bruise. Apply hot wet towels and massage the udder to break up clots. Alternate hot towels with stripping. There is a fine description of this treatment by James Herriot in *All Creatures Great and Small.* A poor and desperate farmer had a cow with such a bad case of mastitis that Herriot expected to find her dead by the next morning. But the man sat by his cow all through the night stripping out milk and applying hot towels (supplied, no doubt, by his equally sleepless wife). In the morning when Herriot returned, the cow was fully recovered. I have never attempted this with a mastitic cow, but I believe the story. It works the same way for human mothers with mastitis, and with this I do have personal experience, having assisted many mothers as a La Leche League leader.

If there is a suckling calf, compel it to suck exclusively from the afflicted quarter. You will need help to accomplish this. The calf will keep quitting and heading for a more rewarding teat because letdown is poor from a mastitic quarter and the cow will kick. One person guides the calf while the other person is on the far side holding the remaining teats out of the way. Nothing beats a calf when it comes to getting out the milk. Mastitic milk won't hurt the calf. If the cow kicks uncontrollably one of you will need to hoist up her tail; this prevents kicking. Do not use the rope cinch method described in chapter 3, "Milking Your Cow," because you don't want to interfere with circulation to her udder.

Don't be swayed by advice in certain veterinary herbals that tell you to cease milking the quarter. The world is full of advice from people who have not had to live with their outcomes. Whatever else you do, keep the milk moving out.

There are reports of success using aloe vera infused directly into the teat as a mastitis treatment. The Louisiana Cooperative Extension arranged for laboratory and clinical tests of aloe vera. In in vitro tests aloe vera did not show bacteriostatic action against *Staphylococcus aureus, Streptococcus uberis,* or *Streptococcus agalactiae.* Infusion of aloe vera into the udder prompted leukocytes to enter the gland. This resulted in a reduction of bacteria present, but no cures resulted. Bacteria numbers returned to previous levels by forty-eight hours after treatment.

There are herbal and nutritional treatments for mastitis. Two that I have used with success on many occasions when mastitis was caught early are cayenne and comfrey. For the cayenne treatment, smear the infected quarter with lard or other heavy grease. Then take a handful of cayenne pepper and rub it into the grease. This primitive poultice is cheap and effective. There are commercial products such as Uddermint that accomplish the same thing more expensively. You are drawing blood to the area. The cayenne burns neither your hands nor the udder, merely warming it. But don't rub your eyes.

Comfrey is a remarkable plant that deserves much praise. I have a huge patch, as does anybody else who lets it loose in a favorable spot. Don't say I haven't warned you. It has a large, juicy leaf. Pick a big handful and make a mush of it in your food processor if you have one; otherwise, chop and pound to release the juice. Rub this mix of leaves and juice on the infected quarter every time you think of it. If you have leaves to spare, feed them to your cow; comfrey is also effective internally. Once they develop a taste for it, cows love it to the point of breaking down fences. It is highly nutritious.

I sometimes resort to an antibiotic infusion with persistent mastitis but combine it with extra milking, massage, and heat, as described.

Once treated with an antibiotic, a cow permanently forfeits her eligibility for federal organic certification. Commercial organic dairies mostly just ship a cow with mastitis to the Golden Arches in the sky. As an uncertified organic farmer, I treat as necessary, withholding milk from human consumption as instructed on the label of the medication. However, by following these suggestions and those described in the mastitis section of chapter 16, I almost never have to treat mastitis.

Prevention is always the best answer. The Iowa State University teaching and research facility in Ames, Iowa, reports having reduced mastitis to negligible levels in their herd of more than 150 cows by making the two following changes in management: They changed to three-times-per-day milking. And they changed from sawdust to straw bedding.

A cow in early lactation presents a very tight, often dripping udder at milking time, particularly if the milking interval exceeds twelve hours. This is an invitation to mastitis due to milk stasis and moist teats. If you can manage three-times-per-day milking for even a short while, chances of mastitis will be greatly reduced.

A number of studies have found that straw bedding remains more sterile than does sawdust. The reason for this is not known. Shredded newspaper is also preferable to sawdust, as is sand.

As noted, the difficulty of treating mastitis without recourse to antibiotics is shown by the fact that dairymen producing organic milk in herds of commercial size often solve the problem by simply shipping the cow. This includes dairies committed to homeopathic treatment.

Milk Fever

This illness is discussed in detail in chapter 15, "Treating Milk Fever." The treatment is IV administration of a calcium preparation and will be the same under all management systems, including organic.

Ketosis

The treatment for ketosis is also the same in all management systems. See the section on ketosis in chapter 16.

Cuts

It is very important to avoid cuts to a cow's teats. These cuts, which are surprisingly common, can lead to minor problems, like unpleasant milkings, to severe problems like mastitis. It is important to try to ascertain the cause of the cut so that future ones may be prevented. If it is a downed fence, for example, it is likely to happen again. Indeed, the vast majority of cuts a cow gets are caused by barbed wire, but they can also be caused by the cow stepping on her own teat as she stands up. This can be a problem for prodigious producers as they age, and may require an udder support or "cow bra" for prevention. Cuts may also be caused by the sharp teeth of a calf.

Nothing works better than pure vitamin E oil on cuts. Always have a bottle of big capsules handy or a little bottle of the pure oil sold with a medicine dropper. If the cut is bleeding copiously, vitamin E won't stick. First make sure there is no dirt under a flap of skin, and do your best to infuse pure vitamin E directly onto the bare flesh. If you are calling the vet to put in stitches, keep applying the vitamin E until he or she arrives. Don't put on any other medications. Try to persuade the vet to let you keep the injury well suffused with vitamin E right up until he or she closes the wound. Then apply it frequently to the stitched incision. Vitamin E works better if not competing with wound powders and salves, all of which interfere with contact of vitamin E with the flesh and don't do as much good.

If you have a comfrey patch, now is a good time to make a poultice of ground leaves and slap it on externally. Comfrey has quite remarkable healing properties. I grind the leaves up to a slurry in the food processor. Cows and any other animals you are treating object far less to the vitamin E treatment and the comfrey poultice than to other medications. I have seen amazing recoveries from lacerations treated with vitamin E alone but will augment it with comfrey if it is available. I keep a supply of pureed comfrey in the freezer for use in winter.

If you are dealing with a teat injury in a lactating cow, you still must milk. Cows will stoically endure being milked by hand, machine, or calf unless the injury is very severe. In the latter event, you can get teat dilators to insert in the teat canal. This allows the milk to drain out during letdown. Maintain conditions as sterile as possible when inserting the dilator. Remove the dilator between milkings. If not using a disposable dilator, sterilize it between uses.

If you're treating a teat or a leg injury, a vet wrap bandage is very helpful. It sticks to itself like Velcro.

CHAPTER 11

Fencing

A good start to the grazing in spring will require a fence the cow can believe in. If she gets started grazing with the knowledge that her pasture has limits, there will be less trouble later when she has explored it and thinks about going farther afield. You could build a permanent fence with posts and rails or boards. This can be a beautiful fence, but expensive to have built by a professional. If you have plenty of trees and time, you could build one yourself. The old-fashioned stone wall worked fine for centuries and still does, but it may be that people used to have stronger backs. Nor are stone walls as simple to build as they look. Skill is involved. But the raw material is free and close at hand in much of the cow-loving world. A cow can jump higher than you would think to look at her. A fifteen-month-old heifer can easily clear four feet. Later in life she may still think she can do it, and that's when she tears her udder. Beyond heiferhood, there are a certain few cows that are incorrigible about jumping fences and will require very convincing fencing at all times. The great majority do not try it unless the fence is damaged and they can more or less push through. But if a cow is being pursued by dogs, or is in heat and hears a bull answering her bellow, or is being separated from her calf, all bets are off.

A basic principle of fence building for cattle that applies to all but stone walls is to have the top board or wire level with the cow's shoulder blade. This is the level at which she carries her nose. Except in the most unusual circumstances, a cow, and other animals as well, is far more likely to breach a fence by going through or under than over it. A cow will scratch her sides on fence posts and stick her head through for choice grass. This is when flabby fencing shows its weakness.

Cows have no aptitude for mounting stone walls. Straight sides three feet high and a broad top will stop them. Yet like every fence on every farm, stone walls too require maintenance. They are such perfect sites for rubbing an itch that a cow may dislodge a capstone. If other stones then tumble, a more nimble cow just might pick her way through to the greener grass she imagines is on the other side.

Barbed Wire

A barbed-wire fence is effective if properly constructed. The materials are cheap, relative to other fencing. A barbed-wire fence is dangerous if poorly constructed because a cow can injure herself if she attempts to go through or over it. The fence is relatively safe if on sturdy posts and stretched good and tight. Four or five wires are needed, preferably five. It is the casual fence that a cow may try to get over.

Electric Fencing

Improvements in electric fencing occur with regularity, so if you want to build one, you will want to consult a supply catalog or your local farm supplier. Electricity is supplied either by a battery pack or by plugging into the domestic electric service. There are also excellent solar-powered models.

The design of both posts and wire varies greatly. Here are some basic considerations.

The traditional style of electric fencing uses smooth wire that runs through insulators attached either to movable stakes or to ordinary fence posts. String and tape with embedded metal filament is now more common but does not have the longevity of wire. It is lightweight and much easier to set up, but any rough handling, including a summer of persistent wind, may ruin it. It is usually white and more visible than wire.

The point is to see that the wire or tape does not touch anything except an insulated surface. If it touches posts or bushes the electricity will head to ground right there rather than continuing down the fence line. To you this may seem obvious, but over decades of hiring helpers, I assure you that understanding the habits of electricity is not an inborn trait. Supervise, or expect to do it over.

The wire must slide freely through the insulators so that you can stand a long distance away and pull the end of the wire to tighten it, like a belt through belt loops. This is important, because wire soon sags and you want to be able to tighten it from afar. Grass growing up under the wire is a constant problem. Animals that respect an electric fence will not graze under it. A powerful charge will singe off any high grass during dry weather. A week of rain, though, and the grass will grow up and short out a sagging fence.

When splicing the wire, double it back on itself and give it at least eight wraps. Repairs will be easier if you are able to slacken and tighten the fence at will.

The most important feature of electric fencing is its portability, so don't allow anybody to wrap wire onto the insulators. Wrapped, crimped wire will end up having to be discarded.

A wire reel makes handling any wire much easier. If you haven't got one, unneeded wire can be wrapped on a board. Keep children well away from the job and wear protective glasses. Flying ends of wire are dangerous to eyes.

One wire about 30 inches above the ground is normally sufficient to retain a grown animal. If there is a calf you will need a second wire closer to the ground. If you're using actual wire, tie surveyor's tape to the wire at frequent intervals so that animals don't inadvertently walk into it. A grazing animal with its head down may get the shock on the back of its neck. The reaction of an animal is to throw up its head and run forward, not dance back. But either way, the fence will probably come down, with the animal loose on the wrong side and singularly reluctant to return by the same route.

As with any fencing, brace the post against the pull of the wire wherever you turn a corner and try for long, straight runs. Where you need a gate, you can get insulated handles that cover the wire and hook onto a loop of hot wire.

Fence testers are available and are cheap. A tester is just a little hook with fourteen inches of insulated wire ending in an exposed tip. Place the hook over the electric fence and ground the tip; if the fence is hot, a little bulb glows red. Tie a big piece of surveyor's tape on the tester and make it a federal offense not to put it back on its peg. Or buy two.

Once a week and following any rain or windstorm, walk the fence line to check for grass reaching it or branches fallen on it. It is my belief that some cows know when the fence has quit.

Before turning out an animal to electric fencing, for the first time, train it by setting a few yards of hot wire across the corner of its existing paddock. Tie flagging on the wire. Watch to see that the animal gets educated. A naive cow is not much impressed by that little wire or tape the first time she sees it but learns fast.

When you hear thunder, disconnect the fence and bring the cow and any other animals in if you can. During an electrical storm an electric fence is dangerous to animals and to buildings. Of course, the cow should come in during thunderstorms no matter what kind of fence she has.

Electric fencing is some trouble but is often the best choice for cow pasture. It is cheaper than permanent fencing, and sometimes you don't even want permanent fencing. Often you can buy an entire electric fencing setup secondhand. Many times somebody will even give miles of it to you. A well-maintained electric fence with two wires will keep out marauding dogs better than any other fence I know of. But see below.

Tension Wire Fencing

This is sometimes called Australian fence. Sturdy metal or wooden posts are required. Six or more wires (light cable) run through staples on the posts and are tightened with ratcheted turnbuckles placed at long intervals. Slotted wooden spacer bars are set onto the wires at intervals between posts to keep the wires equidistant. Each single wire may run for a mile. Tension wire is superb fencing for just about everything except horses. It will hold buffalo and many other wild animals. The problem for horses is that if they paw at it and get their feet stuck, they are likely to panic and get severe wire cuts.

Another disadvantage is that when a wire does fail, it pulls back like a broken tendon and repair is not easy.

A farmer I know with a fine herd of Angus cattle has successfully combined tension wire and electric fence. He replaced every other strand of tension wire with electric fence wire that is served by a supercharger battery pack. Before doing this, he lost calves to neighboring dogs. Since installing this combination-type fencing he has not lost calves.

As you can see, fencing has a role in both keeping animals in and keeping predators out. In traditionally rural areas it has been generally understood that a dog killing livestock has committed a capital offense. Farmers will shoot such a dog. Often they will know the owner and tell him, "Sorry I had to do it, he was after my stock." Or the owner will himself shoot the dog upon receiving the bad news. But traditional farming areas are being swallowed up fast as former urban dwellers buy country property. They are often completely intolerant of the presence of a cow mowing their front lawn for them, finding nothing entertaining in the sight. And they often greet the news that their dog has molested livestock with indignant denial, a state of mind that does not desert them even upon seeing their dog on the carcass, let alone after hearing of your cow's ripped teats. The law in nearly every area is on the side of the cow owner if he or she shoots the dog. But an exurbanite is unlikely to perceive this as justice and tempers are sure to flare. Good fencing before anything happens is wisest.

CHAPTER 12

Making Hay

The hardest part of the work of making hay is getting started. The next hardest part is bringing it into the barn. You might as well make good hay in between.

If the sun shines, your best hay will always be what you make yourself with your own equipment. If you must rely on others to make your hay, they will fit you into their schedule, and you will probably have to wait while they fix something, then it will rain, then the ideal moment for harvesting perfect hay at the right stage of maturity will have passed.

When to Cut

The right stage of grass maturity for haymaking is while it is still green and in bloom or with seed heads barely forming. As soon as heads form, the nutrient value of the leaves declines and rushes into the seed heads. Stems then must get tougher to support the weight of the heads. You need to cut the hay before this process gets very far advanced. Your neighbors may criticize you for making early hay, saying you are losing volume. But the volume and weight they gain is in stems high in lignin, with little feed value and diminished taste appeal.

How to Cut

Before starting to cut, walk the periphery of your field and remove any branches that might foul your equipment and any lumps of last year's hay that got left. You want to be pretty fussy about this. It is ruinous to tempers when equipment is disabled on the first cut. Contract haymakers are so sure you won't have bothered to clean up that they typically start their first cut about four feet into your field. Your field then grows progressively smaller.

If you're mowing with an old-fashioned cutter bar, make your first cut by going all the way around the edge once to open up the field. Cut as close to the fence as you can safely manage. Cut the rest of the field in the opposite direction so you don't have to drive on the standing grass.

Equipment

If you have the fields and are a handy person, haymaking equipment is cheap these days. You might find that over a two- or three-year period the equipment would pay for itself. The bare essentials are a mowing machine and a hay rake. A tedder is also extremely useful. There are other mowing designs you may wish to explorer; all are dangerous. Unless you want to put hay in the barn loose, you will also need a baler.

The older type of mowing machine has a long cutter bar with teeth that slide against each other in a scissors action. It is a design that stood the test of time for a hundred years. The earliest cutter bars were pulled by one or more horses. The action was ground-driven by the wheels. Tractors replaced horses, but the antique ground-driven mower, if you can find one, remains efficient and functional if you drive slowly. Later ones are designed to be operated by the power take-off (PTO) on modern tractors. It is this style that is most likely to be available cheaply now.

A mower requires more horsepower to operate than most other equipment because of the mechanical disadvantage in translating forward motion to crisscross motion. If you find yourself buying equipment, make sure your tractor is powered for the length of the cutter bar.

Be sure when buying any implements that the hitch conforms to your tractor. There are two-point and three-point hitches.

An implement called the mower-conditioner is now in common use. It is cylindrical like a reel mower. It kicks up the mowed grass and crimps it as it passes over the cylinder. This conditioning action is of great value because the bent-up grass dries much more quickly. If grass is mowed in the old-fashioned way with a cutter bar, you really should have that tedder. It has teeth that whip the grass around with an eggbeater action. It is an excellent device, but to use it you must either have a second tractor available or disconnect the mower and connect up the tedder. Two people working the field makes speedy hay. One person mowing and then tedding using the same tractor permits less flexibility because connecting and disconnecting a mowing machine is a nuisance, but not too bad if the weather holds. If you are fortunate enough to have either type of mower plus a second tractor to pull the tedder, somebody can just keep going over the field tedding and the hay will dry very quickly. If the weather is hot and dry with a nice breeze, sometimes it is possible to bale the very same day you have mowed.

The brush hog works on the principle of a rotary lawn mower. Some have a modification that enables them to cut hay and lay it down in a windrow. This method is in general use in Australia (where the brush hog is called a slasher).

Raking

Before the hay can be baled it must be raked into windrows not wider than the baler pickup.

Hay rakes are of several types. The older ones are ground-driven and can be pulled by anything. There is the old-fashioned trip rake, which drags a swatch of hay straight ahead. It has a seat for a rider. About every twenty feet the person on the seat trips the rake and it leaves behind a rope of hay. This is very basic, but the resulting windrows tend to be straggly as you go back and forth over the field trying to trip the rake at the right moment to line up the swatches.

A side delivery rake is another time-honored design. It is superior in efficiency to the trip rake because the action is continuous, so no tripping, but there are more moving parts. A series of teeth is geared to move the raked hay to the side and leave a windrow. The side delivery rake must be driven slowly.

Another style that operates efficiently has a series of big overlapping wheels with teeth that deliver the hay to each other and thence into a windrow quite rapidly.

Rain

If rain threatens, hay should be raked into windrows as quickly as possible. When the sun comes back out, wait until the mowed strips between the windrows dry out, then use the tedder to spread the hay around again so it can dry.

It is critically important that the hay be thoroughly dry before baling. If it isn't, once baled it will mold. It may also get dangerously hot. Spontaneous combustion is possible.

If rain soaks the mown grass while it is still green, and you manage to get it dried out expeditiously, the nutrient value of the hay is very little damaged. If rain hits mown hay that has already dried once and it must be dried a second time, the feed value is greatly reduced. It can still be used to feed winter steers or fed to a cow during the first two weeks of drying off. It can also be fed to a milking cow, suitably augmented with other feed. I have often had to do it, but it is a disappointment all around.

If the hay is rained on and dried out a third time, it is fit only for mulch or bedding. It's still worth baling up and bringing in for this purpose. Take a good look at it and smell it, so you'll recognize such hay in case somebody ever tries to sell you any, misrepresenting it as "only rained on once." If you are fortunate enough to have a brush hog, instead of picking it up, you can go over it and leave it as mulch on the field.

Baling

The world is full of old hay balers, the kind that makes rectangular bales. Somebody is not unlikely to give you one. Most are powered by the PTO (power take-off) on the tractor. They whisk up the hay, pack it into a bale-shaped channel, ram it hard, wrap twine around it, and with an action similar to a sewing machine, tie on the twine . . . except when they jam up or miss the knot. They make a lot of noise. Certain clever people are able to keep them going with impressive consistency. But when they quit, the silence on the hay field is depressing.

If you bale hay, and I really hope you will, use sisal string. Plastic string is a hazard to a cow in case she eats it, and a hazard to the environment. The knotters on older balers will often not tolerate plastic twine, as well.

The baler can be set to the size of bale you prefer. Thirty- or forty-pound bales are heavy enough for most people. Bales stack most efficiently if, like bricks, the length is twice the width.

After the hay is baled, as soon as possible put the bales together in tepees of two, three, or four, so they dry (cure) on all sides. If the weather remains fine, the bales can stand in the field several days. Hay heats up. This way it can do its heating outdoors. If the bales get wet, roll them over so all sides can dry before bringing them in. The hay usually isn't ruined.

Many people prefer to make and use round bales. There are advantages both ways. For one cow, I find the old-fashioned rectangular bale is more versatile and easier to manage.

Hay by Hand

Hay can be made entirely by hand. All the necessary hand implements are available. If you have only about an acre of hay to make, and especially if access is difficult for your or anybody else's power equipment, do consider this. You will need a scythe, long-handled wooden rakes (preferably one for every volunteer), and a couple of real pitchforks with spring steel tines. A manure fork won't work. A scythe must be exactly proportioned for the man who mows, as perfectly sized as a golf club. There will be an adjustment on the handle for at least one of the handholds. The blade must be razor sharp. A man can then march forward rhythmically swinging his scythe, hear birdsong, and smell only new-mown hay as he lays it down. Everybody else can follow along and fluff up the hay, exactly as in a painting by Constable. When the hay is dry, rake it into windrows, then divide the windrows into a series of haycocks to be pitched onto the wagon. Like bales, a haycock can take a little rain without serious damage. Few women are tall enough or strong enough to use a scythe, although for some of us anything is possible.

What my boys and I often have done is to mow the hay with a mowing machine, run the tedder, rake it into windrows and haycocks, and then pitch it loose onto the wagon. Everybody can have a nice hayride back to the barn. Then it gets pitched loose into the haymow. Using this method we have filled the barn with hay from fifteen or twenty acres in a season.

When loading hay onto a wagon, fill up the corners and sides before filling in the middle. Keep stomping down the middle as you build up the sides. Otherwise the load will be unstable. Do this also in the haymow, including the stomping. Unstable hay can slide like an avalanche while at the same time making it difficult to get a forkload out when you want it. There is a nice description of the skill needed to make a load of hay in "The Death of the Hired Man" by Robert Frost.

I have described three ways to make hay, and more variations are possible. It is clear enough why farm couples have traditionally been pleased to have large families. On farms, children were and are an economic asset. If you have a small family, try to have lots of friends. Haymaking is a lot of work, but wonderfully worthwhile. Make it a party. Serve them haymaker's switchel, a traditional power drink made of whey enlivened with a splash of apple cider vinegar and a handful of fresh raspberries.

Putting Hay in the Barn

When stacking either baled or loose hay, start out with a plan and appoint yourself field marshal to make sure others do it your way. Otherwise, enthusiastic helpers will often have your hay unloaded and into the barn while you're still mopping your brow.

If your hay storage area has a dirt floor, lay down boards or pallets to keep the hay from getting damp. Ideally, leave a draft area under the hay that's high enough so the cats can hunt under it. Otherwise it's a free zone for rodents. If you must stack directly on the ground, lay the first layer of bales on edge so the strings don't rot. If the hay is of varying quality, keep the types separate and don't allow your best or worst hay to be totally buried and unavailable. You'll want to be able to access your hay according to your needs without a lot of restacking.

Pile the bales like bricks, making the Flemish bond. The subsequent rows need to overlap and be tied into each other by setting them in opposite directions. Cover the designated floor area layer by layer rather than building walls of bales. Alternatively, pave the entire area with bales lined up the same way; then place the next layer running crosswise. This latter method of stacking is best reserved for a large delivery of uniform hay, or else your varying grades will be sandwiched and be inaccessible. Careful stacking will prevent an avalanche of hay descending on children playing or onto yourself as the stack diminishes.

Those who deliver hay will rarely stack it for you, but if they do, speed, not orderliness, is their priority.

Loose hay also must be put in the mow in an orderly fashion, as noted. Keep filling the corners and sides of the room so the height at the edges stays just ahead of the middle until you level it off at last. You will be feeding it out from the top layer in a pattern opposite to the way you put it in the barn. The order in which it was put in is the *only* way it can be forked loose, so you will appreciate that you do not want yourself or others having to climb up to dismantle some hay Matterhorn created by somebody trying to see how high and far he could pitch hay.

If not thoroughly dry, both freshly made baled and loose hay can get surprisingly hot after it is put in the barn. The dampest parts get the hottest, which is another good reason to get it good and dry in the field. Until you are experienced at making hay, it's a very good idea to check the hay before you go to bed. Stick your arm down among the bales, and perhaps crack one open. Reach into a layer of loose hay and note the temperature. If you are at all worried by what you find, set the alarm and check it later in the night. No small number of barns have burned due to spontaneous combustion occurring in hay that wasn't put in dry enough. Even when it doesn't get dangerously hot, too much heat in the hay lowers feed quality and reduces palatability. When you find moldy hay, you know it was put up insufficiently dry.

Children large and small are attracted to playing in the hay. One hates to deny them this pleasure and usually won't be able to make an injunction stick in any case. Look the place over and try to get rid of every hazard you can think of. Be *especially* particular that pitchforks are hung up after every use. If the barn has a hay drop for feeding to a lower level, fence it off and keep it closed. Make sure the area down below is kept clear and not used to park lawn mowers. Be fanatical about barn safety.

When feeding out hay, sometimes you will run into bales that are mostly good but have moldy streaks. Rather than discarding the entire bale, consider feeding it outdoors spread on the ground. The cow can pick it over and mold spores will drift away. Hay dust and mold are almost as hazardous to a cow as to a horse.

Testing for Moisture Content

The moisture content of hay can be determined using a microwave and an accurate scale that measures in grams.

Collect a sample of hay from the windrow or bale. Snip it into two- to three-inch lengths.

Find a microwave-safe ceramic plate or heavy paper plate. Weigh the plate and record the weight, in grams, or adjust the scale for the tare of the plate.

Place a glass three-quarters full of water in the back corner of the microwave to prevent charring and to protect the oven.

Weigh out 100 grams of snipped forage. Spread it evenly on the plate. Heat in the microwave for two minutes. Remove the plate from the oven, weigh the plate of hay, and record the weight.

Mix around the hay, rotate the plate, and heat an additional thirty seconds. Remove the plate, weigh again, and record. Repeat until the weight does not decrease by more than two grams per thirty-second round in the microwave, recording the weight each time. The final weight is the dry weight. If the hay starts to char before this point, use the last recorded weight.

To figure the moisture content, subtract the final dry weight from the initial weight. Divide this difference by the initial fresh weight and multiply by 100.

Hay is ready to rake up at 35 to 40 percent moisture. Bale at 20 to 25 percent.

This method was developed by George Haenlein at the University of Delaware.

Maintaining the Hay Field

Hay takes everything away and puts nothing back on the land. Haying or cropping far more than grazing is responsible for rundown land in marginal farming districts everywhere. Grazing at an appropriate stocking rate takes about two-thirds of the grass, leaving the remainder to carry on photosynthesis and restore many nutrients. Manure left behind by animals adds more nutrients. The hoofs of livestock tread in manure and dead herbage, which builds topsoil. If the land is organically managed, armies of dung beetles will carry the manure into the soil. Areas with a moderate stocking rate and plenty of opportunity for rest and renewal by snow and glacial meltwater remain stable indefinitely. Consider the Swiss Alps, where cattle and goats have grazed for centuries.

Not so with the hay field. Haymaking leaves nothing behind but roots and three inches of stem. If you're haymaking, you need a consistent program for nourishing the hay field. A three- or four-year rotation involving a legume such as alfalfa, soybeans, or peas followed by row crops is a traditional approach. Soil preparation for planting can be combined with manuring. Trash from the crops can be disked in to add more organic material. A three- or four-year rotation that includes using hay land as pasture can be highly beneficial to both uses. Among other benefits, it will starve out parasites and discourage weeds.

For those of us without farm equipment or manpower to operate it, some other approach to maintaining fertility must be found. If you are able to arrange for the spreading of manure from a nearby hatchery or layer operation, and you and your neighbors can deal with a few days of noxious odor once a year, this will do a good job. Not all poultry waste is acceptable under organic standards. Municipal sludge has its advocates, but I am not among them. It smells if anything worse than hen dressing and contains viable seed. According to its source, it may contain heavy metals. It will surely contain pesticides.

Resting the field for a year by brush-hogging in mid- or late summer after ground-nesting birds have fledged will create valuable mulch. If you can't make hay or elect not to, it is a mistake to simply leave the grass standing. A lot of that dead grass will still be there next year and will interfere with haymaking. Also, perennial weeds and tree seedlings will get a start.

If it isn't possible to fertilize the entire hay field, it is worth doing just a strip. Next year the enhanced appearance and production of the manured strip will be an inspiration for further improvement.

One of the easiest and cheapest aids to fertility on most soils is simply liming. Lime isn't toxic, but you should get a soil test to determine need and application rate. There are cheap and simple devices to pull with a lawn tractor, push like a stroller, or even hang over your shoulder and carry to distribute lime and any other granulated fertilizer you choose to use.

Fertility can also be added by aerial spraying. An ordinary garden hose can be fitted with a nozzle and plastic canister. Fish emulsion or any soluble fertilizer can be added this way. If you have experience with maintaining a healthy lawn, it is a short walk to maintaining a healthy hay field. On my place I have a dramatic example of the difference fertility makes to grass. My vegetable garden is separated only by barbed wire from a piece of neglected land used for hay and pasture. That grass is sparse and weedy. On "my" side of the fence there is a manure pile and some intensively cared-for plots. Here the very same grass, with the same amount of rainfall, grows almost four feet tall and is bright green and juicy looking. An armload I can scarcely carry can be cut from a thirty-six-inch square. On the unimproved side of the fence a thirty-six-inch square would just about fill a dishpan. This explains why it is impossible to state exactly how much land it takes to support a cow, either for her pasture or to provide for her winter hay.

If you do not make hay you will have to buy it, thus adding the fertility from somebody else's land to yours. Their mistake. Your gain.

Raw Milk and Whey as Fertilizer

A source of fertility uniquely adapted to cow owners employs raw milk or whey. It is useful on both large areas or the smallest flowerpot with results that are consistently impressive. For years I had been congratulating myself on the exuberant growth of my rose bushes. I treated them with whey and milk washings and thought myself quite original, but no, some farmers have known this for years. The concept received a boost when David Wetzel, a retired corporate CEO, and Terry Gompert, a USDA Extension agent serving as skilled observer, received recognition for the impressive grass growth on Wetzel's fields. Wetzel's retirement project had been to start a dairy farm and cheese plant. Burdened with excess skim milk and whey, he dressed sections of land with various ratios of raw skim milk or whey and water using a tractor-mounted spray rig. Results have been so gratifying that Wetzel urges those needing fertilizer to order up a tanker of milk from the nearest dairy. He suggests paying the farmer a premium over the meager price he would get from the co-op. "It's worth more on the land," he declares.

Many formulations are possible, but according to Wetzel, a mixture of three gallons of milk to seventeen gallons of water is highly successful not only to promote vigorous growth of grass but to build lasting fertility.

Who has milk to pour on the ground? It seems we all do from time to time. If you do not sell milk and do not have a calf, pigs, or poultry to keep up with the supply, you too will often find yourself saying, "What am I supposed to do with all this milk?"

Skim milk may be better than whole milk. Too much butterfat may clog equipment. So far as I can tell, whey works as well as milk. Even small amounts are worth using. Whenever I have whey I dilute it three-to-one and anoint whichever plant is calling out to me, either watering the ground or sprinkling it on the foliage. The results are uniformly rewarding.

Many fine soil amendments exist. I am not sure that anything tops composted cow manure. However, as a low-cost high-powered fertilizer, milk has few rivals, and I know of none as easily and pleasantly managed. Unlike the case for other fertilizers, animals can return to grazing immediately after a field has been treated with milk or whey.

In recent years I have made it a practice to rinse out every jar or glass that has contained milk and pour the washings into a handy bucket. I use this cloudy water on potted plants. The results have been hugely rewarding. Effortless Boston ferns and orchids grace my winter windows, and amazingly, I now am able to winter over fuchsias that not only stay in bloom but are free of aphids. Milk is amazing.

Although my own pastures are nothing to brag about, they have shown steady improvement. Clarinda has little cause to complain. She and her calf have a twenty-acre field and know where the good spots are. She is fourteen years old and still has a calf every year, freshening at five gallons a day, an impossibility for an undernourished cow. She is twice the age of the average cow in a commercial herd.

We once visited the showcase garden of a celebrated organic gardener who lives in Maine. He and his wife said they planned to get a cow but were waiting until they had built up their soil fertility for her. They had that exactly backward. When you get a cow, she builds up the soil fertility for you.

Their famous vegetables didn't look a bit better than mine.

CHAPTER 13

Pasture Management

Whether your cow is on a total grazing program or you wish only to optimize a small pasture area, many of the same principles apply.

If you have a grazing area of even one acre it is possible, with active management, to provide the major portion of your cow's livelihood for as many months as the grass will grow. A good regular soaking with a sprinkling system will greatly extend the grazing. As noted earlier, the ideal grass length is four to seven inches. Dividing the field with electric fence until the cow has eaten about two-thirds of the grass in her assigned area is best. Then you can mow it before seed heads form and it gets stemmy. Many species of grass will come back thicker than before.

Coburn Farm, pasture home of my cows Clarinda and Helen
Photograph courtesy of Joann S. Grohman

Mowing

Mowing improves the grazing in a pasture almost immediately. If seed heads have already formed, oddly enough, for a couple of days a cow will often eat that headed-up grass after it is mowed, even though she ignored the standing crop. A mowed field gets busy and grows again as soon as it gets watered or rained on. But it grows back faster if mowed before seed heads are fully formed. Like any plant, the ambition of grass is to propagate; once heads have fully formed it takes it a while to get back in the mood to start over.

Use a brush hog, reel mower, sickle-bar mower, or a mower-conditioner ("haybine"). Do not use a rotary or flail mower on a pasture as it throws the cow pats all around. Left undisturbed, the pats form what parasitologists call a "circle of repugnance." The grass around each undisturbed pat will form a bright green hummock that a cow will ignore until the third year, thus avoiding reingesting parasites. This results in the characteristically hummocky appearance of cow pastures in long use. If you prefer your ground to stay flat and you have enough pasture so that you can afford to fence your cow off part of the land in a three-year rotation, brush-hogging will smooth it out. Brush-hogging has other benefits as well. It discourages perennial weeds (as does mowing), and it gets rid of tree seedlings that are always trying to reclaim the land around my place.

Another fine option for cleaning the land is to run poultry on it. The "chicken tractor" concept is highly effective. A little chicken hut with a fence framed around it like a big playpen can be dragged forward daily to the benefit of all. No parasite survives the busy hen. Even simpler, allow the chickens to free-range. My self-renewing flock of about fifty assorted birds has kept the nearby acres free of all sorts of annoyances. I have not seen a tick in years. They keep down flies too.

Weeds

There are few perennial weeds that a cow will eat. A great many of them are toxic or have stickers. Tree saplings are a curse.

A method of weed and sapling control that also promotes optimal pasture utilization, once common and now being revived, is to keep mixed species of animals. Cattle, goats, sheep, pigs, and poultry can all share the same pasture if desired, so you can consider that for the future. They have differing tastes in herbage and are host to different parasites, and their droppings contain different nutrients, too.

There are some plants that no animal will eat. These include goldenrod and buttercup.

Mowing broadleaf plants before seed heads form will eventually exhaust them without recourse to herbicides. But the presence of some forbs is desirable. Some provide nutrients otherwise low in grass and may even be sought by your cow. It has been found that many forbs foster more efficient rumen fermentation, with the result that less methane is produced. Nonetheless, a large weed population is indicative of low fertility. This pasture needs feeding.

Pasture Improvement

You will wish to take some measures to encourage good grass and discourage weeds. Almost any grass is good and doubtless will support a cow. Whatever came up in your pasture is what likes to grow in your neighborhood. Many fine grasses can be seeded to provide balance and variety, but go slow on special mixtures. Often they embody more theory than practice.

There are some pasture grass problems for which we all need to be on the alert. Extension Service recommendations for seeding of new grass varieties over past decades has resulted in introduction of exotic species. It takes experience to recognize the various species of grass. If there are farmers in your area, find one and ask him or her if there are undesirable grasses for which you need to be on the lookout. The Extension Service has free educational materials on grasses.

Fescue is a grass introduced to the United States from New Zealand. When tall, it may contain toxic amounts of an endophytic mold. Tall fescue is unpalatable but drought resistant and may be the only forage left standing.

Johnson grass, arrow grass, Sudan grass, and common sorghum are all capable of causing cyanide poisoning under certain circumstances. Stress to these plants increases toxicity; drought, trampling, and freezing are common stresses. Nitrogen fertilizers and spraying with herbicides also increase toxicity. All types of wild cherry carry the threat of cyanide poisoning, but cattle avoid wild cherry unless nothing else is available. My fields are surrounded by it, and none of my animals ever touches it.

Sweet clover, as hay or silage, that becomes spoiled may contain toxic levels of dicoumerol, a blood-thinning agent induced by a mold. This is not a problem when grazing fresh clover.

Bracken fern is seriously toxic. Cattle do not eat it if other forage is available, but it may get mowed along with hay. It does not lose its toxicity in storage and may be difficult for a cow to avoid in her manger. If bracken or other questionable weeds have been baled into your hay, fluffing up the flakes of baled hay will make it much easier for your cow to skip over what she doesn't fancy.

Wild cherry, bracken fern, and other toxic plants are to be avoided. But the grasses remain important forage despite the problems they may present under certain conditions of climate or stage of growth. Cattle not stressed by hunger are remarkably good at avoiding toxic plants.

If your field is not too rocky or remote for heavy equipment, grass seed can be direct-seeded by slicing into the turf without plowing it up. This requires special equipment, but in many areas this work can be contracted for. Your extension agent may recommend this procedure, but keep in mind: he or she did not go to school to learn how to tell you to do nothing. Neither does your agricultural supply store have anything to be gained by leaving your field alone. Certainly don't buy seed before soil testing. Soil amendments, or a dose of some nutrient, may be all that is necessary to permit the grass species you already have to compete successfully against the weeds. You may just need to add more water. An oscillating lawn sprinkler often does wonders.

Mowing and leaving the grass as mulch, plus the manure left by grazing animals, will help keep pasture in good heart. A layer of snow each winter adds nitrogen. All the same, a cow is taking away the nutrients she uses to make milk, and more. Plan to periodically add back fertility by spreading manure and lime. A yearly soil test and consultation with experts in your state will soon make you an expert too.

Maintaining Pasture

Once established, your pasture needs to be kept mowed or grazed and fertilized in order to be permanent. It will not die unless there is severe drought or overgrazing. Freezing may occasionally damage it but, if you have native grasses, seldom seriously. Native grasses are adapted to prevailing climatic conditions.

If you are faced with a barren, plowed-up waste, you will then almost certainly have to seed it; otherwise it will be colonized by weeds. If it formerly had a broad-leaf crop on it, it may even have been treated with an herbicide specific to grass. You could then investigate the reputation of alfalfa in your neighborhood. Being a broad-leafed plant, it would not have been affected. If the land formerly had corn on it (corn is a grass), a broadleaf herbicide may have been used, in which case the clover contained in most grass mixtures will not grow, nor would alfalfa.

If you are planting either alfalfa or clover, remember that these two fine legumes can cause bloat. If this is to be your cow's only pasture, you will need to plan on buying some hay to feed your cow before she goes out in the morning, which means keeping her in at night.

If you are contending with a plowed-up area, this means it is accessible to heavy equipment. It may be easiest to contract out the necessary disking and seeding. If you must proceed from a standing start, a balanced chemical fertilizer may be your best option, unless you have time to get in rye and then plow it under as green manure. Because of the drastic decline in farming, you may be unable to find equipment, labor, or real animal manure with which to fertilize. This might be a good time to try fertilizing with milk (see page 193).

When you are seeking advice on choice of seed, make it clear whether the land is to be used for grazing, hay, or both; mixtures are designed for each use. Even so-called permanent pasture will benefit from reseeding about every four years if you are not satisfied with what is coming up by itself.

Keep the cow off the part you intend for hay, or there will be dung in the hay and trampled spots. A hay field comes up every year just like pasture or lawn. If you don't fertilize it, you will get a little less every year. In the fall spread lime if needed, hen dressing, or just about anything you can find to augment your own cow's supply.

Comfrey

Comfrey is a perennial leafy plant with deep taproots. It is hardy and is widely fed to cows in Russia. It contains as much as 24 percent protein and is rich in calcium and other nutrients. In favorable conditions it grows prolifically. Over the past hundred years it has fallen in and out of favor as feed. Its detractors note that it contains pyrrolizidine alkaloids capable of causing liver damage when consumed in large quantities. To put things in perspective, this same anti-nutrient is found at approximately the same levels in spinach and chard.

There exists some confusion as to which cultivar deserves the Latin name *Symphytum officinale* L. The kind I have fits the description of Russian comfrey. It exhibits lush, dense growth and easily crowds out competitors. It propagates by both seeds and roots. By the third year the taproot may be three inches in diameter and over three feet long. It is brittle and will break if pulled, soon to be sending up new shoots from the fragments. It also sends out a lateral root or arm that claims new ground. It grows three or more feet tall and has a fuzzy stem with small, pretty blue flowers and rough leaves. If you plant it you need to be pretty sure you want it. It took me several years to stop fighting it and learn to love it. Once they have developed a taste for it, my cows, if seeing me in the garden, will hurry from across the pasture in hopes I will cut an armful for them. As forage it is said to be possible to feed comfrey for up to 10 percent of a cow's roughage. I have attempted to dry it for winter use, but it shatters hopelessly. It would need to be bagged.

Comfrey has an impressive ability to improve soil. If you succeed in reclaiming its ground either by hand-digging or by solarization (covering it with a tarp and letting the sun cook it to submission), the new ground is light and wonderfully fertile.

As a healing herb comfrey has few equals. I make a slurry of it to rub on an inflamed udder or any type of wound or injury, bovine or human.

Woodland Grazing

If all you have is woodland, you can put your cow right into that. She can help you clear land by eating all the small stuff and the tops of suitable deciduous trees such as willow and alder. Once the land is opened to the sun, grass will come up by itself and stumps will slowly rot away. This is what the early settlers did when they had no fields. They couldn't go out and buy bales of hay to augment a cow's diet as we are able to.

In colonial times hay was so dear that salt marshes were intensely sought after. They are natural stands of grass and totally free of weeds; colonial plats show salt marshes crosshatched with lines attesting to the importance of ownership of holdings as small as a quarter of an acre.

Milk production will be poor if a cow has nothing but woodland in which to make a living. You will need to bring her some hay. If you do this at fixed times she will wait for it, eat it, leave, and get back to work in the woods.

Woodland grazing carries inherent risks, among which is the possibility that in the absence of grass a cow may ingest poisonous plants. Some of the most common are as follows.

Poison Hemlock (*Conium maculatum*), Water Hemlock (*Cicuta douglasii* and *Oenanthe crocata*)

Throughout the Northeast, poison hemlock and water hemlock are a hazard to cattle grazing in damp places in woodlands or along their edges. Best to learn to recognize these and similar plants and try to either eradicate them or fence them off. Both poison hemlock and water hemlock resemble carrot tops and Queen Anne's lace. Water hemlock is the most violently toxic plant in the United States. It is adapted to moist sites, grows two to three feet high, and is palatable. Poison hemlock may grow ten to twelve feet high. It is unpalatable, and cattle are unlikely to eat it unless other green forage is unavailable.

White Snakeroot (*Ageratina altissima*)

This plant greatly resembles the garden plant ageratum. The flowers are small, white, and fluffy. It is found primarily in the Midwest, growing one to three feet tall. It is primarily a woodland plant but persists in cleared areas. It is best eradicated by pulling it up. Cattle should be fenced away from areas where it is found. Early symptoms of poisoning include trembling and loss of appetite. One large dose of the plant may produce poisoning. It also has a cumulative effect, so that persistent small amounts are equally dangerous. The milk of affected animals can be fatal to calves, lambs, and humans. Meat from poisoned animals is also toxic.

White snakeroot poisoning is relatively rare, but any cow owner should learn to recognize the plant as its range may be spreading.

For a comprehensive listing with full descriptions of poisonous plants, see the website of the USDA Agricultural Research Service.

Predators

Unlike goats and sheep, a cow is rarely attacked by predators, but bring her in during hunting season. Packs of domestic dogs are a threat to all livestock. Although most of my animal-keeping experience has been adjacent to huge forests, it is only to dogs that I have lost livestock.

Keep a bell on your cow, especially if she has access to woodland. Then you will always know where she is. Should she be chased, her bell may alert you.

CHAPTER 14

Housing, Water Systems, and Supplies

A cow's housing requirements are flexible. In a balmy climate she may need only shade, preferably an airy place with dim light where she can escape from flies while she chews her cud. In a cold climate a weatherproof building of some sort is required. Cattle are remarkably hardy. But a cow cannot endure steady cold rain and wind and must have somewhere to retreat from these. Dairy cattle endure freezing temperatures better than heat so long as they have constant access to hay and adequate water; rumen fermentation produces a lot of heat. They suffer if required to stand in icy mud. Although a cow endures dry cold well, there is an added metabolic cost, as more of her feed must go to keeping herself warm. An unheated draft-free building is what you want. You'll have a happier cow and go through less hay. I have kept a cow in a suburban garage very successfully. A garage is in many ways ideal.

Building a nice little barn can be a lot of fun. If you're starting a barn from scratch, here are some suggestions that will add to your future convenience and pleasure:

Plan for accessibility by a truck and trailer for ease of hay delivery and in case you ever need to load your cow.

If you have a hillside site, it's wonderfully convenient to be able to pitch manure down to a lower-level run-in area and directly into a manure wagon.

Water and electricity to the barn aren't essential, but they make life a lot easier. If you are in a cold climate get reliable advice about putting in a water line that won't freeze. Place electric sockets where there is no chance a cow can stretch her neck to lick them. She can stretch a long way.

Put grain storage near the door so you don't have far to carry it. Fence or wall off the grain storage area so that the cow cannot possibly get into it. Very important: put a convenient latch or door closer on the grain room so that it is easy and convenient to *always* close it after yourself. This rule will prevent tragedy. Overeating of grain, particularly chicken feed, can be fatal to cattle, horses, and goats due to impaction of the gut and to bloat.

The barn floor can be concrete or wood. New concrete needs to cure for some months; during this period the area where the cow stands needs to be covered with bedding or a cow mat. Unless scored, a concrete floor will always be slippery; if it is scored, it is best cleaned with a pressure hose.

You will need a loafing area for the cow and a separate area where she comes to be milked, preferably with a stanchion. For generations cows have been kept locked in stanchions, some never getting out for exercise. Many still are. Cows do in fact adapt to this life remarkably well, but it ruins their feet. It seems sadly confining for a family cow. Even if you don't have fields, I think it is important to have something in the way of an exercise yard. My cow walks in and out at will. At milking time I let her into a separate milking area. This makes it easy to keep the milking area clean.

The milking area, besides a tie-up or stanchion, needs a sturdy low shelf where a bucket can be set safe from spilling. You might also want a spring-type scale to hang from the ceiling, with a hook on which you can hang the bucket to weigh the milk. While certainly not necessary, weighing your cow's production makes possible a useful record.

A hay storage area above the cow is ideal. It is wonderful insulation from heat and cold and easy to throw down. Our very best barn ever had three levels; the cow was on the middle level. Manure collection was down below and the hay above. This was possible because both the two lower levels were stepped into a slope. The hay level was accessed by a little bridge from the hillside.

Upper-level hay storage can also be achieved with a hay elevator. These are technically simple and usually run off a small electric motor. It is basically a ladder with a continuous chain running up the middle. There are teeth on the chain, and a bale of hay set on the chain just rides upstairs. Snap it up quickly if you ever see one cheap. A hay elevator makes your friends a lot more willing to help fill the barn.

A cow needs at least fifteen gallons of water a day. A high producer will need considerably more in hot weather. Don't slight this requirement in planning your facility. Water should be constantly available if possible. Failing that, someone will need to offer water at least four times a day. If the water isn't constantly available, milk production will suffer. Restricting a cow's water will also restrict her eating because she cannot continue to process her food without water.

If you are in a cold climate, life for you and your cow will be easier if her living area is insulated well enough to prevent her water from freezing. If this is not feasible (as it is not here in Maine), a good option is a submersible water heater. I find that except in the coldest weather, the heater does not need to be on continuously. Tubs with built-in wiring are also available. These are smaller but safer.

We have insulated housing around the hose bib. The hose stays rolled up inside it when not in use. A light bulb keeps the hose and faucet from freezing.

In areas of mild climate, the water problem is easily solved. A float valve that clamps on the edge of the water tub and connects to a garden hose will do the job. Such a float valve has a hose connection, and everything can be connected in the same way a sprinkler or nozzle connects to the end of a hose.

An upright pail is not sufficient for a cow's water; it will not hold enough and will tip over. As discussed in chapter 9, a 50-gallon stock water tank is a good choice.

Bedding

Sand, straw, spoiled hay, sawdust, or wood shavings are among your choices for bedding. If you have your cow living in a box stall, sawdust or shavings are most easily managed because they are more absorbent than hay or straw. But use whatever is plentiful and cheap. Another factor to consider in your choice of bedding is that straw or hay will make better compost for your garden. Sawdust and shavings require great amounts of nitrogen to decompose, which means that much of the nitrogen in the manure may be used up in decomposing the sawdust instead of being added to the soil of the garden. It also takes longer for sawdust to decompose than it does for hay or straw. However, once decomposition is complete, the sawdust-manure mixture still has considerable value for plants. Sawdust has been associated with a higher incidence of mastitis than other bedding but should not be a problem if kept dry.

Hay is not economical for bedding unless it has been spoiled in the making or is an excess for which there is no market. If you do use hay and plan to put it on your garden, you will want to make sure it is thoroughly composted so that grass seed is no longer viable. When bedding with straw or hay it is easiest to keep adding a new layer and build deep bedding. This is what we do; we remove the worst of it every day and keep it dry by adding waste hay. The cow remains warm and dry. This is a fine system with one caveat: the area must be accessible to a front-end loader for yearly clean-out.

The deep bedding method does not work at all with sawdust or shavings, since a thick mat of these materials soon becomes fouled, and skimming off the surface doesn't result in a clean bed. The big advantage of sawdust or shavings is that they are easy to handle.

If you are in a grain-producing area you probably can get straw at a reasonable price, and this is the bedding of choice because of its superior fertilizer value.

Ventilation

It is worse for a cow to be too hot than to be too cold. A cow can live in subfreezing housing without difficulty (although a portion of her feed will go toward keeping her warm), but she will be uncomfortable and her milk production will drop if she is too hot. (In the hot summers of the interior valleys of California, for example, dairy cows drop in milk production, even though lush feed is plentiful.) Worse than heat is lack of fresh air. Your cow will be processing at least fifteen gallons of water and up to seventy-five pounds of feed each day. She will require huge volumes of air in doing this. In a closed building she can use up the oxygen. That is why in a dairy barn full of cows, you usually see the windows open and ventilation fans running even in subfreezing weather.

Cows require ventilation, but not drafts. Ventilation should be provided by windows high up in the wall, so that drafts are above the cow. If you are building new housing or choosing a corner of a barn for your cow, try to choose a side away from the prevailing wind for ventilation windows.

A Milking Stanchion

Your milking area should be comfortable for you and your cow. The stress of learning to milk will be much reduced by having your cow's movements restricted. Your cow will hold still with a minimum of forward and backward stepping if you put her in a stanchion for milking. A stanchion is designed to hold her head in place – taking advantage of the fact that she has a large head and a narrow neck. You can build a good one out of 2x4s. The structure can span a space of any convenient width, from a minimum of four feet. Run a 2x4 horizontally about twelve inches up from the ground or floor, and another 2x4 about four and a half feet above the floor. Fasten these securely. Run vertical 2x4s across the space on ten-inch centers, except for

Photograph courtesy of Max Luick

the middle space, where your cow is to put her head through. Make this a fifteen-inch opening for the present. Now add a second 2x4 parallel to each of the two horizontal 2x4s, making a sandwich filling of the vertical 2x4s. Cut one more 2x4, a foot longer than the space between the two horizontal bars. This 2x4 will be centered vertically in the fifteen-inch head opening. Drill through the lower 2x4s and this last 2x4 to make a swivel point for this piece, which is used to lock the cow in the stanchion. Bolt through the three 2x4s at the bottom, and double-nut the bolt so it is loose enough to allow the closing piece to move back and forth easily. Put your cow in the stanchion with her head through the center opening. Adjust the movable 2x4 to a comfortable position and mark the spot. Drill through at the top and arrange a pin to hold the closing piece in place at the top. A stop block will make it easier to put the pin through the three layers of 2x4s. Once she walks into the stanchion and puts her head in the grain pan, close up the pivoting 2x4, put the bolt through the hole, and if it's correctly designed, that's where she'll stay until you say otherwise.

If you are in a dairying area, secondhand steel stanchions can almost certainly be found. Some prefer to add a barrier on the far (non-milking) side of the cow such that she cannot inch away during milking. Your cow may like to lean on this barrier during milking so it should be well-braced.

The floor where your cow stands to be milked can be concrete, gravel, wood, or plain dirt. Concrete is not necessarily more sanitary. If the cow is not to slip, the surface must be roughened and will then require hosing off. Wood is my favorite. It scrapes clean, it is a comfortable surface for standing, and it dries quickly. I put a little pile of shavings on the floor behind my cow in case her barn manners should fail her.

Supplies

In your cow house you should have the following items to assist you in the care and management of your cow:

- A curry comb to groom your cow; a once-over on the milking side before sitting down prevents debris falling into the milk bucket
- A small bucket and cloths for washing your cow's udder before milking
- Paper towels to dry your cow's udder
- A feeding pan for your cow's feed ration or treats
- A grain scoop (a two-pound coffee can will do) to measure out the dairy feed
- A three-legged milking stool, a sawed-off chair, or any suitable stool about twelve inches high
- A manure fork with six or eight tines

Occasionally an infrared heat lamp is helpful, especially with a young calf. If you have a place to plug in a sturdy cord with a lamp socket, you are well prepared. For greatest safety, have the heat lamp wired through a permanently mounted porcelain socket, and be certain that the wiring is heavy enough to carry the additional wattage. Cattle are very sensitive to electric shock. They can be killed by voltages that humans can withstand. Do your wiring properly, with exposed cable in conduits or out of reach and well fastened. Avoid dangling extension cords that could be chewed. Cows have a long reach.

The Hay Feeder

To eliminate waste, use an arrangement that requires the cow to put her head through a barrier to eat. If the space the cow puts her head through is V shaped, she won't find it so easy to take a mouthful of hay and pull it through, wasting it on the floor. If she must take a little trouble to get her head in, she will stay there and eat for a while. The hay feeder should be at ground level, and it needs a floor. Hay feeders placed high up inside a box stall promote waste. With every mouthful the cow eats, she pulls down several more and treads on them.

Hay nets as used by horses cannot be used by cows. Unlike horses, cows do not have upper incisors. They grab food with their tongues, which does not work for eating from a hay net.

A hay ring purposed for a round bale also works fine for baled hay.

Photograph courtesy of Max Luick

CHAPTER 15

Treating Milk Fever

Calcium tetany and parturient paresis are more descriptive names for milk fever, a hormonal disorder that may occur in high-producing cows just before or soon after calving or (rarely) at other times. There is no fever. It is a form of paralysis brought on by elevated calcium demands at the onset of lactation. Its principal victims are high-producing cows and goats, although it is not unknown in sows, dogs, and cats. Among cattle, it is more common in Jerseys than other breeds, because Jerseys give more milk in proportion to their size.

Steady calcium levels are essential to muscle function; calcium blood levels must be maintained as precisely as those of oxygen or glucose. A drop quickly becomes critical, and so a complex system of hormonal controls exists in all animals. Where the sudden calcium demands of lactation onset are exceptionally high, as in the higher-producing cow, the system may prove inadequate. The resulting paralysis is called milk fever. No diet can be counted on to meet the calcium demand. The hormonal control system is designed to pick up calcium from bone reserves. But during the cow's dry period, when calcium demands are low, the system "goes to sleep."

The milk fever prevention diet described in chapter 5 is designed to keep the cow's system slightly starved for calcium so that the needs of the unborn calf cannot be met by the cow's diet alone and thus bone mobilization remains active. Vitamin D should also be given. Milk fever usually occurs only after a normal calving. The stress of difficult calving seems to activate the adrenal and other glands so that the vital hormone levels are in place when needed.

Symptoms

The very first symptom is an unsteady gait. You may notice it during the day. Soon the cow lies down, and if you feel her ears they will be cold and usually droopy. After calving I check my cow's ears hourly and have been known to sleep in the barn. If the cow is on her feet, early symptoms include paddling with the hind feet and swaying as if she is about to fall over. Once down, she will twist her head and neck to the side as if there were a kink in her neck. Her nose becomes dry. Another warning signal is if your cow, which was bright and active caring for her newborn calf, becomes listless and inattentive. This could be a symptom of any malaise, but at this particular moment in your cow's life, suspect milk fever.

If not treated, the symptoms increase in severity until the cow goes down, she begins to go cold (you can feel the ears, legs, and other extremities becoming cold first), her head remains pulled around, and her pulse slows. She no longer recognizes you. All rumen activity ceases. After several hours, she will be unable to hold her head up and will fall over flat, gasping for air. If nothing is done, she will die, usually of strangulation by fluids in the windpipe or rumen pressure on her heart. It isn't normal for a cow to lie out flat, and the body fluids and pressures go wrong in this position. The only time you see it other than in acute illness is for brief periods in the field on a warm, sunny day, and then only by an immature animal. If you ever see your cow lying flat on her side, run to her calling for help as you go. Pick up her head and do what you can to tip her up onto her brisket.

Treatment

Milk fever rarely strikes a first calver. But any dairy cow should be monitored every few hours from the time she starts calving onward, through the day after calving. If you see the described symptoms, act. Milk fever does not go away by itself. The longer you wait, the more severe the treatment must be and the greater the chance of losing your cow. Call the veterinarian. I find that not all veterinarians react in the same way to milk fever. If yours has a large animal practice, meaning that he specializes in cattle and horses, he will think nothing of injecting the necessary calcium gluconate (or calcium magnesium gluconate) solution into your cow's vein. If he is also the type to show up promptly, I would summon him and let him undertake the treatment. Some large animal vets have a pretty laissez faire attitude, because they see so many cows and don't realize how much your cow means to you. They may arrive late and play down the need for treatment. The small animal man may suggest you drive to his dispensary and pick up a calcium tablet, or if he comes he may suffuse the injection under the skin rather than into the vein. He is certainly unaccustomed to cows and may be afraid of them.

Either way, you are well advised to be prepared to provide treatment yourself. Giving the calcium solution under the skin is not difficult, and if given early it is usually effective. There is no hazard in it for the cow, and it is good insurance. If the treatment is much delayed, though, she may rally after the treatment and get up but then go down again. Treated again, she may repeat the recovery only to go down once more. This has been termed "the downer cow syndrome." The cow doesn't appear to have anything wrong with her, but she does not get up, or if she does she does not stay up, no matter how many times you give her calcium. She won't eat much of anything and she just gets thinner and thinner, a lingering death. I have had more experience with milk fever than anybody would ever want, and despite what anyone says, I am a firm believer in injecting calcium directly into the vein if the case is at all advanced. There are varying opinions as to how long it takes for calcium injected under the skin to take effect, ranging from twenty minutes to four hours. The latter appears to be more correct, and this is four precious hours lost when things are going from bad to worse before the turnaround begins. The dangers of injecting calcium in the vein or artery are infection and the possibility of shock to the heart if cold fluid is infused in the blood rapidly. I have had good luck with the procedure and no casualties caused by it.

Equipment

If you want to be prepared to treat milk fever yourself, you will need some simple equipment available at farm supply stores. The equipment consists of a flutter valve, which is a one-quarter-inch rubber tube with a rubber cap on one end to fit the top of the bottle of calcium solution. The other end of the valve slips over a standard hypodermic needle. A refinement is to have a section of glass or plastic pipe in the middle to enable you to see the fluid pass through the tube, but it isn't necessary. For subcutaneous injection you will need a needle with a bore or lumen that is about one-sixteenth of an inch. Special needles made for cows have a wide collar around the top to aid your grip in forcing the point through the skin, which, after all, is cowhide. If the special needle isn't available, any large-bore needle will do. Needles that are one and a half to two inches long are best. A needle with a smaller lumen is used for injecting into the vein; the large needle would allow the fluid to pass into the bloodstream too rapidly. The lumen of the smaller needle should be half the diameter of the large-bore needle.

Finally, you'll need a supply of the life-restorative fluid. It is sold under various names, but the common variety will have a solution of calcium borogluconate at 20 to 40 percent, with magnesium, phosphorus, and dextrose added. There is also a straight calcium borogluconate available. The mix, sometimes called MPD (magnesium-phosphorus-dextrose), is useful because it covers any possible magnesium or phosphorus deficiency and adds a pick-me-up in the form of the dextrose. Calcium-magnesium solutions and equipment for injection can be purchased in agricultural supply stores in some states. They can also be obtained from a veterinarian or online from an agricultural supply house (see the list of sources at the end of this book).

The calcium solution is not expensive. For one cow, keep two or more 500-ml bottles on hand. Once you have seen this equipment in use, you will derive considerable peace of mind in having it ready to use if required. A short story in a national women's magazine told of a family that moved to the country and bought a cow. The story had a tragic ending: the cow died in calving, obviously due to milk fever, although neither the author nor the family appeared to recognize this. You need not be caught in such a tragedy.

Subcutaneous Injection

If you've decided that your cow needs a calcium injection, proceed without delay. First, sterilize your equipment (the flutter valve and needles) by immersing them in boiling water for a few moments.

Take a clean bucket, fill it partway with hot water, and set the bottle of calcium solution in it. This will warm the fluid to somewhere near body temperature, and it provides a place to put all the bits and pieces. Otherwise, you are almost sure to lose the needle once you have completed the injection, providing yourself with the classic opportunity to look for a needle in a haystack.

If your cow is able to move, put her in a box stall if you have one. Confining her will cause her to stand relatively still while you treat her. Otherwise, once poked with a needle, she may give you a merry chase, forgetting her condition for a while. If your cow is in an advanced case of staggering and swaying, do your best to get her into a stall with good bedding (she will need sand underneath any bedding on a concrete floor in order to get up). At least get her under cover if at all possible. Don't leave her locked in a stanchion, since she may hurt herself if she goes down. Remove all buckets, feeders, and anything she might fall on. If she has calved, remove the calf. (I had a calf squashed flat when her mother went over on her with milk fever.) A sick cow might stay down if you push her over, which would be useful for your first attempt at an injection. If she doesn't, tie her head closely to something firm while giving the injection, and get someone to help you by holding the equipment with one hand and the cow's tail to steady her with the other.

The subcutaneous injection can be made in the skin of the shoulder. You will be sticking in the large-bore needle by itself, no tube attached. Go back a spread hand's distance from the front of the shoulder and the same distance down from the backbone. Push the needle in at a sharp angle to go under the skin, not into the muscle. Pick up the skin with your hand to feel the thickness of the skin and fat layer. You can hold the skin with one hand and push the needle in with the other. Push hard. See that the needle end is free under the skin by moving it slightly. That accomplished, the needle should remain there while you take off the top of the calcium bottle and replace it with the rubber cap of the flutter valve.

Hold the bottle upside down in the air, and as soon as the fluid emerges from the far end of the tube, slide the tube end onto the top of the needle. The fluid should start bubbling in the bottle and flowing through the needle. Plan to inject half of the bottle at this site. Massage the skin downward and away from the needle to disperse the fluid under the skin. This is optional but speeds the effectiveness of the injection and prevents the development of a lump under the skin. If you cannot spread the fluid very well, it probably means it has pocketed in the tissues, and it will be best to stop now and inject the remaining contents of the bottle on the other side of the shoulder. This at least will keep the size of the lump small, and usually it will disperse and be absorbed.

When the bottle is empty or you are finished with that injection site, pull out the needle, rub the area, and put everything in the bucket to be washed. If not washed thoroughly, the tube and needle will be clogged by the sticky dextrose.

Leave your cow in a comfortable state. Prop her up with hay or straw bales so she cannot roll over flat. If possible, arrange them around her like building blocks so she hasn't space to roll over (unlike a horse, a cow cannot roll over and may die if she does). Untie the cow, or else during clumsy attempts to arise she may stumble and fall with her neck outstretched, causing her to strangle. Do your best to keep her from collapsing in a corner. A cow rises by lurching forward; if she gets herself wedged in a corner, it will take six strong men to shift her back into the middle of the room before she can even try to get up. Someone should watch her closely until the calcium takes effect.

Intravenous Injection

If your cow is in a more desperate state and you decide to inject the calcium directly into her bloodstream, here is the procedure. An inexperienced person will require a helper. You also need a wristwatch. The bottle of calcium solution needs to be at or close to blood heat – which is very similar to that of humans – by setting it in a bucket of warm water. You use the same equipment as for the subcutaneous injection, except that you will need the smaller-bore needle. Another useful item is a piece of small-diameter rope several feet long.

Your cow will already be down if you have decided this procedure is necessary. Prop her up with hay bales if at all possible. Try to get her right up on her brisket. This is not at all essential to the injection, but it is to her survival, since a cow left lying on her side will bloat (blow up like a balloon) and cut off her own air and circulation. Put every effort into sitting your cow up, for example by prying her up with a 2x4, but if she is too far gone to make it possible, get on with the injection without delay.

Have your helper hold the cow's head up and around to one side. The effect is to stretch out straight and tight the lower side of the neck, where the blood vessels are located. You can work on either side of the neck, since the injection is effective in either the jugular vein or the artery. Work on the outside curve of the neck, since that position pulls the vessel tight. In practice, you work on what you can get at, since cows tend to move themselves into a corner in their efforts to get up after going down with milk fever. You probably can see the skid marks in the bedding where her feet have been working to get up.

With the lower side of the neck stretched tight, feel for the throb of the vein. It runs next to the windpipe, which is a stiffer tube. The vein will have the feel of quarter-inch tubing running along the lower neck, an inch or so from the fold of the neck. If you cannot feel it, have your assistant hold the head higher and turned more sharply. Your assistant will become very tired. If you have a snap nose ring, you can tie the cow's head in place, but I always feel so sorry for the cow that I want to hold her head in my arms. If you continue to be unable to feel the vein, use the small-diameter rope mentioned above. Put it around the cow's neck a short way behind the head and tighten it up to pinch the vein, which makes it easier for you to feel the throbbing and pumping of the blood at the point where the rope crosses the blood vessel.

Swab the site with alcohol. Pushing the needle through the skin layers and then into the blood vessel is usually a two-step process. Get the needle through the skin at the proper level to intersect the vein. Then, using one hand to hold the needle end and skin folds in line, jab into the blood vessel. If you have scored a hit, blood will drip from the needle. Now remove the tourniquet. Here is where a second helper is very handy. You need to have the calcium bottle upside down in one hand, with the flutter valve attached, its tubing coming down full of fluid and dripping. You then slip the tubing end onto the needle, hold the bottle low, and let the calcium drip very slowly into the bloodstream. The trouble is that if you haven't got this all ready and waiting at some convenient and safe place (which there seldom is), you will have to abandon your carefully placed needle to get the calcium ready and you'll usually lose the vein in the process. Your assistant can't let go of the cow's head, either, lest the needle be dislodged. The only thing I have thought of, other than a three-person team, is to hang the calcium bottle from a beam, at what one hopes turns out to be a convenient height, and to put a clothespin on the dangling tube, so that it's ready to be transferred to the needle.

Before connecting the tube to the needle, have it running freely so that you do not inject air into the vein. You can judge the speed of flow by the bubbling of air in the bottle. Take fifteen to twenty minutes to administer the entire bottle. Administering the solution slowly is important. If the calcium enters the bloodstream too fast, the cow could die of shock. If there is a disturbance of the needle, you can verify that it remains in the vein by removing the tube for a moment and seeing that blood flows out of the needle.

When the injection is completed, just pull out the needle. Usually there will be no significant bleeding. Because of the outward flow of the blood, the infection danger is minimal if the needle was clean. See that your cow is comfortable. You might want to have a little hay and a bucket of water handy for when she comes around. Sometimes you will see brightening in ten minutes, and sometimes a cow will even get up in that time. Often she will begin to brighten but will not arise for several hours.

The first signs of brightening are swallowing, licking her nose with her tongue, and belching. She will also start to look around. If it is the middle of the night you can go back to bed for a while, but check her again soon. Then, if she is not up, and hasn't been up (you can tell by her position and the presence of fresh dung), do your best to make her get up. A cow may decide that things didn't go well the last time she was up and will just lie there. For her own good, she must get up. She cannot function well lying down, and her udder will get into trouble as well. But of course you can't even begin to lift her. Shout at her sharply. Clap your hands in her ears. Slap her on the rump. Once she starts to get up, help her steady herself. A firm grip with both hands at the base of the tail will help hold her back end in line while she gets up in front. If she has been down a while, her legs will have gone to sleep. This is a terribly difficult problem with a cow that stays down any length of time. If your cow has brightened up and shows signs of recovery, she should be able to get up.

If doubts persist as to the cow's condition, she may require further calcium treatment. I often use an injection under the skin after an injection in the bloodstream, as a slower-acting, longer-lasting backup treatment. If the cow is down, or luck is not with you in getting the needle into a vein, don't be discouraged. Keep trying. One night when it was –25°F and we could not get a vet, my daughter and I struggled on until dawn trying to hit a vein in our cow Faith. I was never sure whether our failure was due to our own ineptitude or the fact that the cow seemed to have virtually no pulse to show us where her blood vessels were. Faith obviously suffered each time I jabbed her. We never did get it right, but she lived. I decided later that the adrenaline response to our painful jabbing probably kept her going until an earlier subcutaneous injection could take over. I think Faith understood we were trying to help her. She changed from an aloof cow to one that was truly friendly toward us both from that day forward.

In another undeserved success on a cold winter's night when my cow Jasmine was down, I gave up and injected the drip into milk veins on her abdomen. It worked and she lived, but both injection sites exhibited big lumps that did not resolve for many weeks.

Calcium Paste

Calcium paste is a supportive follow-up treatment given by mouth. It comes in a tube that fits into an ejecting holder exactly like a caulking gun. It may also be used as a precautionary measure before symptoms of milk fever appear. A number of formulations exist. Most claim to be "less caustic," which tells you something about how it tastes. The paste does a pretty good job of providing calcium to a cow that is on her feet and showing appetite. To get it down her throat you will need to have a halter on your cow and her head snubbed up short. Try to have a helper. Get the tip into the side of her mouth in front of the molars and squirt it in. You can administer it several times. Follow it up with treats. I suggest warm molasses water to wash it down. The paste is capable of giving a cow a sore throat, which is a very bad thing indeed right now, when you want her to be eating as well as possible. I avoid the paste if I can.

Milk fever is indeed a serious illness, and treatment, although usually dramatically successful, is demanding. But do not worry unduly. I have never lost a cow from it when I was actually present. The only tragedy occurred once when our cow was left unattended while I was away. Learning that she was ill, my young son and I hurried home, but by the time we arrived forty-five minutes later, she had slipped and rolled into a puddle from which we were not strong enough to move her. The actual cause of death was bloat or suffocation due to her position. Had she not been left unattended, undoubtedly she would have survived.

Recovery

If your cow comes around easily, no special treatment should be required afterward. As soon as she is steady on her feet, you can let her calf come back to her under supervision. Feed your cow well. Tempt her often with high-quality roughage and grain in small amounts. The idea that it is better to restrict feed to avoid recurrence of milk fever (on the theory that the more she eats, the more milk she will produce and the more calcium will be used up) may possibly be helpful in milk fever but can lead to acetonemia (ketosis), which is a more difficult metabolic disorder.

A traditional approach to avoiding milk fever recurrence is to go easy on the milking for one or two days. That is, don't take all the milk when you milk. The theory is that by reducing the amount of milk removed, you will reduce the amount produced and thus reduce the demand for calcium. Studies have shown that this does not necessarily forestall a relapse, but it does predispose to mastitis. One of my cows permanently lost production in one quarter when I overdid this method.

An effective compromise might be to milk out enough to relieve excessive pressure and do this frequently enough to keep bacteria from gaining a foothold. Save all the colostrum you collect and freeze any not needed. If your cow seems to want it, let her drink some of the colostrum. It can be life-saving. Give her anything she will eat. A cow that does not eat will soon have ketosis, the first symptom of which is loss of appetite.

CHAPTER 16

Diseases and Disorders

Most cows most of the time are perfectly healthy. There are only a few diseases or disorders that I have personally encountered during my many decades of experience. If you encounter any with your family cow, they will probably be the same ones: mastitis, milk fever, torn teat, and bloat. Even with the most conscientious management it is likely that you will meet them at least once.

Observe Your Cow

Cattle are weather-hardy compared to goats or horses. Jerseys are considered the most weather-hardy of the short-haired breeds. Given reasonable shelter in high winds or icy rain, they are unlikely to catch cold. Respiratory problems are much more likely when your cow is confined in a humid atmosphere with little ventilation or must sleep where she cannot avoid drafts. Your cow will need a dry place to lie down and chew her cud. She will come to the gate and ask to come in when continuous cold rain spoils all the fun of being out, if she does not have free access to her shelter. She is not likely to become ill if left out but she will produce less milk, in part because she will use a lot of her energy to keep warm and in part because once she gets the idea of coming in stuck in her brain, she may not go back to grazing for some time.

A cow will communicate to you how she feels if you stop and look at her carefully every day. When you see bright eyes, a nose covered with dewdrops, a shining coat, and a reasonably full rumen, you know all is well. If any of these signs is absent and your cow appears not to care about anything, she is in trouble. Jerseys especially care about things intensely. Cows see to it that the proprieties are observed; if there are two cows, one must make it clear that she is boss over the other, no matter how much trouble it takes. If your cow quits caring about all this, she is ill.

You need not worry about your cow all the time. Just look at her carefully at least once each day. Pay extra attention around calving time. Watch carefully for mastitis when she is being dried off, when you no longer have clues from the milk filter to alert you. The most important part of cow health is feeding. Feed your cow well, on clean, high-quality hay and ample grazing. See that she has adequate minerals by using a supplement or a mixed dairy feed containing added minerals. In most climates extra vitamins in winter are valuable. And a reliable source of clean water is essential. These matters are discussed in chapter 9, "Feeding Your Cow."

You will probably wish to get a veterinary manual for in-depth discussion of common cow diseases and disorders. Here is an outline of the most common problems in cow health. Since you will encounter disorders not by name but rather by symptoms, I will list them in that way.

Lameness

The most common causes of lameness are overgrown hoofs and foot rot. Another possibility is a mechanical injury caused by the cow running into something, especially going around a tight corner, or catching her toe in a crack. This will usually right itself. A badly torn toe will profit from veterinary attention. The vet can glue on a lift, which takes the weight off the injured toe. For overgrown hoofs, have a veterinarian or farrier trim the hoofs. You can learn to trim hooves yourself, but if an overgrown hoof has gone unattended long enough to make your cow lame, it is best to have it done by an experienced person the first time. Even an experienced person will, in his or her zeal to get results, often trim too close and draw blood, which can prolong the lameness. For reasons not well understood, some cows' hoofs grow very fast and need a lot of trimming, while others wear and grow at about the same rate and stay nice. Fast-growing hoofs are a nuisance, because cows are not as well adapted to cooperate with hoof work as horses are. Hoof trimming tends to be an adventure no matter how many times the cow is trimmed. Professional hoof trimmers with an amazing lift and tip contraption are available in many areas. The ground and pastureland, which naturally wears the cow's hoofs as she walks will influence the frequency that is needed, but trimming will generally be needed at least once a year.

Foot rot is an infection of the foot, sometimes called foul of the foot, or other regional names. The most prevalent cause is the organism *Fusiformis necrophorus*, which, if your new cow should be infected, was undoubtedly brought with her from her former home. Wet, slushy, dirty conditions predispose to the disease, especially where the feet could also be wounded as on rough concrete (concrete is a problem either way – smooth concrete is dangerous because it is slippery). If your cow is lame, check her feet by lifting them and cleaning them out. A hose can be useful for getting them clean. Usually you can see evidence of foot rot in the form of a lesion or swelling, with a vile-smelling pus. Leave the foot clean and paint it over with an antiseptic. Even an old-fashioned household disinfectant will do. The infection usually responds dramatically to an injection of a sulfonamide or antibiotic. Although these are available without prescription, you probably will want to ask your veterinarian to come to treat this disease. When foot rot is a chronic problem, control is achieved by making a foot bath through which the cow must walk regularly. Solutions of copper sulfate or formalin at 2 to 5 percent are used in the bath.

An older cow can get arthritis. Make sure she has a comfortable bed free from drafts.

Inability to Stand: The Downer Cow

At calving, the most likely cause is milk fever, which is discussed in chapter 15. Other possible causes are grass tetany (hypomagnesemia) and poisoning. Grass tetany occurs only when the cow is on lush pasture that is deficient in magnesium. The cow will first be seen acting strangely in the field, with convulsive movements and staggering. (The disease is sometimes called "staggers.") The cow will go down, her temperature will rise, and she will die unless treated. The treatment is injection of a magnesium solution under the skin, as is done with a calcium solution for milk fever. The response in the case of hypomagnesemia treatment is not as quick as with milk fever and so, if left until too late, can fail to save the cow.

Sometimes a cow suffers nerve damage during calving or dislocates a hip by falling. There are various types of lifts that can be used to get a cow up with the aid of a front-end loader. Often the cow will eat and drink normally even while down. If it is impossible to get her up, you must roll her to her other side a couple of times a day or her weight will cut off her circulation. It can be a terrible situation. Yet I have known of beloved cows cared for for many weeks that finally stood up and survived.

Poisoning

Poisoning can occur when certain weeds or tree foliages are eaten, especially when the cow is in a pasture where she has not been before. Generally, if there is adequate food, a cow will not eat poisonous weeds. There may be certain dangerous plants in your area that can be controlled. It would be a good idea to ask your extension agent and nearby livestock farmers about this. In some areas ergot is a danger on certain plants. Yew tree foliage is extremely dangerous.

A common source of poisoning is lead. It is most often found in old paint on surfaces your cow can chew or lick. Lead tends to have a sweet taste that attracts the cow, or she may go after it because of a mineral deficiency. Lead is a cumulative poison, but it doesn't take much. My vet warns that only a few flakes may kill a cow. Among useful remedies is the calcium borogluconate used for milk fever.

Another possible poison source is chemicals put on plants to control bugs and weeds. A cow may be grazing in or near garden areas where a former resident has used these poisons. One of my readers suffered a tragedy with her cow because insecticide had been dumped on her field by a neighbor. It had occurred before she and her husband bought the place. The grass came up poisoned.

See also the discussion of poisonous plants in chapter 13, "Pasture Management."

Cough and Pneumonia

We're speaking here not of the belching of methane after eating concentrate feeds, but of a persistent cough while the cow is resting or grazing. The most likely causes are pneumonia and hardware disease. Additional symptoms of pneumonia will be high temperature, labored breathing, nasal discharge, and a general state of depression. Watch the breathing. If it is quick and shallow with a jerk to it, there is certainly respiratory trouble, the most common cause being pneumonia. It is effectively treated with antibiotics. In the event of pneumonia, there must be an immediate change of the conditions that brought on the disease. Stuffy, humid quarters with damp bedding are the most likely culprits. A heat lamp is a good investment and, although usually used for calves, will be much appreciated in drying out and keeping warm the ill cow. Be sure there is plenty of ventilation (but not drafts). And of course be doubly sure the lamp is not in a position to ignite dry hay. Flunixin (trade name Banamine) will help her feel better and perhaps get her eating.

If treating pneumonia organically, supportive treatment as above along with close nursing care is called for. Warm molasses water with lots of vitamin C will help. If she will eat, fresh fruits and vegetables, and comfrey if you have it, will help your cow fight her illness.

Hardware disease ("ironmongery disease" is the British term for it) is caused by the cow inadvertently picking up a nail or some other metal object along with her feed. Such an object usually lodges at the front of the second stomach (reticulum) and then may move forward through the diaphragm and pierce the heart. Or it could just lie there and do nothing. One veterinary source tells of the removal of twenty-six nails, a piece of barbed wire, a schoolboy's compass, and two pennies from one cow. In an examination of 4,400 carcasses at a meatpacking plant, residual lesions caused by foreign objects were found in 70 percent of the animals. Do not provide the opportunity for hardware disease to develop. Be absolutely fanatic about the care and use of objects such as nails, wire, and string in any area your cow will ever visit. Cows do not go about picking up these things on purpose, but they are unable to separate them from their feed, especially if they don't see them. Have you ever looked for a nail in a haystack? When you are driving nails in the cow barn, don't go about with a handful of them; pick them up one at a time. When pulling an old nail, hold it with the other hand. It can easily fly into the hay and be nearly impossible to find. And when you do lose a nail, hunt for it until you find it, using a magnet if necessary.

The symptoms of hardware disease vary, depending on where the object goes. If your cow coughs and you can't figure out why, consider the possibility that she has swallowed a foreign object. A small, smooth magnet can be purchased at farm supply stores and put down your cow's throat to collect and hold pieces of ferrous metal that are swallowed. The metal will stay on the magnet during the remainder of the cow's lifespan, thus preventing it from perforating the rumen. I generally put a magnet in my cows by the time they are a year old, as a preventive measure. If you butcher on the farm, you can retrieve the magnet, if you are especially enthusiastic.

Plastic hay string is also a serious hazard. If you have to buy hay tied with plastic string rather than sisal (which eventually digests), be just as fussy with it as you are with nails. Plastic bags are a menace as well. They smell like food, and cows often eat them. Once inside the cow they fill with fluid and occupy the space for food or lodge between stomachs with fatal results. Teach everyone on the farm to pick up all plastic bags. Be serious about this.

Eye Discharge

The three most likely causes of a continuous running from the eyes are pink eye, vitamin A deficiency, and a foreign object. Ophthalmia, or pink eye, is caused by an organism rapidly spread to all animals by flies. It is for the most part a warm-weather disease. Flies can carry the disease from one farm to another. A copious discharge starts about two days after the fly infects the eye. If treated without delay, pink eye is not a great problem. If left untreated, it can cause blindness. It is treated by applying ointment to the eye. While someone holds your cow's head, you can lift an eyelid and squeeze in some ointment. There are quite a number of different products for this, but basically what you need is a corticosteroid and antibiotic in a small tube. Once the tube is opened, keep it clean, and discard it after the current trouble is cured.

Many people have reported success using some of the cow's own milk in the infected eye.

A piece of straw or grass seed, such as from foxtail, lodged in an eye can cause a lot of misery. Sometimes it will not clear without help. In this case, get a helper to hold the cow's head. Lift the eyelid by the lashes and wash out the object with a clean, dripping wet soft cloth or piece of cotton. If the eye still runs, the pink eye treatment may clear any infection.

Vitamin A deficiency is actually a possible factor in any infectious condition. If you use a commercial mixed dairy feed, you are unlikely to have this problem because vitamin A is usually included in the feed. Your cow will also get vitamin A from green hay in winter and plenty of it from the grass during the grazing season. If your winter feed is poor, brown hay, and you don't have any other vitamin A source in your cow's diet, a deficiency is probably the cause of runny eyes. You may also observe night blindness. If you can get cod liver oil in bulk, it is an excellent source of the vitamin. Green-colored, sun-dried hay is also excellent. Any chronic infection deplete the body's stores of vitamin A.

Eyes Sunken

This symptom accompanies any prolonged and debilitating illness in cows, just as it does in humans.

Eyes Protruding

Some Jerseys seem to have somewhat "popping" eyes. My vet thinks it is a characteristic of occasional members of the breed and is not associated with any health problems. However, exophthalmic goiter occasionally occurs in cattle. It is due to chronic iodine deficiency.

Nasal Discharge or Slobbering

The most common causes, other than an object stuck in the throat, are hypomagnesemia, pneumonia, and hardware disease – all three mentioned earlier in listings of their symptoms.

Scouring (Diarrhea)

Most of the probable causes of loose, runny manure have to do with diet, including:

- Lush green feed
- Nitrate poisoning
- Sudden change of feed
- Faulty ration
- Poisonous plants

This discussion applies to adult cows only; calf scouring is a somewhat different thing and is treated in chapter 7.

When cows go into a new field of lush spring grass, you can expect their hindquarters to be a mess. Counteract this by offering some hay (which your cow may not eat, because she is enjoying the grass too much). Cut back on the mixed dairy feed when your cow goes out to grass in the spring. And watch for bloat.

Heavy applications of commercial fertilizer that are not thoroughly dispersed can cause nitrate poisoning. Antidotes for this have been developed so that the fertilizer applications can continue, but the best thing to do is go organic and stay away from petroleum-derived fertilizer. Unfortunately in some farming areas the water supplies periodically reach toxic levels of nitrates. Yearly testing of wells is advised.

A faulty ration can have a few different problems. Moldy hay or silage, for example, will irritate the digestive system and cause gastroenteritis. Change your feed promptly. A ration heavy on finely ground wheat or barley will send the cow's digestive system into a tailspin. If you make your own dairy feed ration, grind the grain coarsely or roll it. Rolled oats or rolled barley are very good as a base for a ration.

A sudden change in feed can also bring on scours; see page 227 for further discussion.

When all of the above have been eliminated as the cause of scours, consider the possibility of poisoning.

As for all the symptoms discussed in this chapter, I list only the most likely causes. There are many more. If none of the ones listed appears to be the answer, seek the help of your veterinarian. Scouring (with wasting) is the key symptom of Johne's disease (caused by *Mycobacterium paratuberculosis*). Your cow won't spontaneously develop this disease. She has to get it from another cow. Unless she had it when you bought her or you introduced a new cow, you can dismiss Johne's. It is not readily contagious.

Loss of Appetite

Loss of appetite could be a symptom in many disorders, but one condition has it as its only initial symptom: ketosis, also known as acetonemia (see below).

Ketosis or Acetonemia

The name *acetonemia* refers to the smell of acetone – as in nail polish remover – on the cow's breath. Some describe the breath as smelling of overripe pears. *Ketosis* refers to the breakdown products of fat metabolism (ketone bodies), which increase greatly in the bloodstream as a cow attempts to meet the high energy demands characteristic of the onset of lactation by mobilizing her fat reserves. Ketosis might also be thought of as hypoglycemia (low blood sugar), since this is in fact what is inducing the ketosis. Cows have relatively small stores of glycogen. Other tissues can get along briefly without glycogen, but the brain must be fed; the brain is able to fuel itself with ketones at least temporarily.

Treatment involves frequent feedings of tempting high-sugar foods. You may have to open the cow's mouth and push it in if her appetite is really gone. Push some choice hay in, too, or the tidbits won't digest properly. Another possible treatment is to give her a drench of molasses with equal parts of warm water or apple cider vinegar. One reader in desperation used maple syrup and reported good results. Raise her head up no higher than her shoulder to avoid getting the solution into her windpipe. Stop after the first mouthful to make sure it is going down properly. The object of drenching rather than merely encouraging the cow to drink from a bucket is to bypass the rumen and send the sugar where more of it will reach the gut intact and restore glycogen in the liver. But first drench with three tablespoons of bicarbonate of soda in six ounces of water. If possible, let this run down her throat in a more normal drinking position, as its purpose is to temporarily reduce rumen acidity so that any sugar entering it is more likely to reach the next stomach intact. A third common treatment is sodium propionate, also given as a drench.

Glucose can be given as an injection.

Try to keep the cow moving around, as exercise is essential to proper digestion and to combat the constipation that always accompanies acetonemia. Frequent feeding is necessary to help keep blood sugar raised. Devoted cow owners often spend hours each day hand-feeding cut-up apples, carrots, or pumpkin and handfuls of hay and clover until the cow finally recovers her appetite.

Prevention is what we should aim for, as this disorder causes poor milk production and sometimes becomes chronic or even fatal. Although most often associated with early lactation, ketosis can occur at any time that energy demand exceeds energy intake. Don't let the cow get too fat during the dry period, as this induces fat burning. Don't let her go very long without food in front of her during the calf's rapid growth period at the end of pregnancy, and never let her be without food in early lactation. Be particular that she has fresh water. See that she gets some exercise, and don't subject her to stresses that interfere with eating or digestion, such as shipping or taking away her companions.

Acetonemia (ketosis) can also occur as a side effect of infection with its attendant loss of appetite; mastitis or metritus from retained placenta could be behind it. While fresh, high producers are its most likely victims, it is not unknown in dry cows in late pregnancy and in heifers. With early detection and your attentive care, it is highly unlikely that it will become dangerously advanced.

Fever

After you've become acquainted with your cow, you will be able to suspect high temperature, even without using a thermometer. Extra-warm ears and udder are indications. Taking a cow's temperature is not especially difficult. A veterinary thermometer has the same markings as one intended for human use but has a ring on the end to attach a string. You can contrive one from an ordinary thermometer with adhesive tape and string if you like. You will need the string in order to retrieve it should it sink from sight. The normal rectal temperature is 101°F or slightly above. Just lift the tail, put the thermometer in slowly, and try to keep your cow from moving around for a couple of minutes. She may present you with a cow pat on the first try and you'll have to start over. If the temperature is above 102°F, look for the trouble. Several diseases already mentioned can be the cause, including pneumonia, hardware disease, hypomagnesemia, and some kinds of poisoning, but the one to suspect first is mastitis.

Mastitis

Mastitis usually affects only one quarter at a time. If your cow is in milk, you'll be aware of mastitis right away because of hardness in one of her quarters that is still present after milking and a flakiness or stringiness in the milk. You can monitor for mastitis daily with virtually no trouble simply by looking at the milk filter after straining the milk. If there are little blobs or strings, your cow has a touch of mastitis. This is subclinical; there is little in the way of symptoms other than the little blobs on the milk filter. These aren't significant unless there are a lot of them and they appear day after day. Subclinical mastitis will nonetheless sap your cow's strength and damage her udder. Active mastitis usually results also in milk that moves sluggishly through the filter. If your warm milk is reluctant to strain, the cow has mastitis.

Another foolproof daily monitor for mastitis is to taste a little from each teat before you milk. It should be sweet and delicious. If it is flat or salty, a problem is brewing. There is no health risk to you in this practice. It has the advantage of alerting you immediately to trouble and enabling you to apply topical treatment before you leave the barn.

Mastitis is painful. Active mastitis in a lactating cow will cause her to kick during milking. It will be difficult to milk her out completely and you will get less from the affected quarter. The quarter will remain hard after milking.

Dry Cow Mastitis

If the cow is dry, you won't have the milk to monitor. The times to keep a close check for dry cow mastitis are right after drying off and again as calving approaches. As a general principle it is unwise to take squirts of milk from a nonlactating cow, as this disturbs the antiseptic plug that has formed in the teat orifice. But if you suspect mastitis, break this rule. What comes out may be thick, but it should not be stringy and horrible. It may be somewhat translucent and not white like milk. This of itself is not abnormal.

Since it may be difficult to evaluate the implications of what you squeeze out just before calving, temper your judgment with what you can feel in the udder. If there is a hard spot (not just firm but hard, like a baseball) inside an otherwise fairly soft udder, it is mastitis. You had better milk out all you can from the affected quarter, and then, if the hardness is still definitely there, and if there is any fever, you or your veterinarian may want to inject a short-lasting antibiotic, such as penicillin-streptomycin in combination, in a muscle, to quickly knock back the infection, even though it is a shame to have the calf get a big dose of antibiotic with its colostrum.

Treatment

It is illegal to sell milk from a cow being treated with antibiotics. The preparation used will state the number of milkings for which the milk must be discarded. If you have pigs, you may feed the milk to them; milk also makes outstanding fertilizer. If I am not in a position to devote hours to curing mastitis with milking and massage (another useful thing is hot compresses), then I use antibiotics. I use them infused in the affected quarter of the udder and give an intramuscular injection as well, since the tubes squeezed up the teat canals (this is easier to do than it sounds) probably don't reach all of the infected tissue.

Half a dose of antibiotic is no good. Since the body's natural defenses are impaired by the infection, the dose must be adequate to knock it out until the body can take over. In a moderate case of mastitis, one good dose may be sufficient. In a more serious one, you had better figure on several days of treatment and on drinking a lot of water while you wait for your milk to be good again. I haven't given exact doses, because they vary with the antibiotics used and also because I think it is best to get the aid of a veterinarian, at least until you have more experience. With hand-milking you are less likely to see mastitis. Don't let these pages frighten you. If it is not far advanced, mastitis, although a nuisance for all, in most cases is easily treated.

See chapter 10, "Your Organic Cow," for a discussion of alternative mastitis treatments.

Vaginal Discharges

When your cow comes in heat, there is usually a discharge of clear mucus. This is a normal sign of heat and is helpful in letting you know the ideal time in the heat cycle for insemination. If the discharge is bloody, it is too late for insemination to be effective. A little clear mucus is seen from time to time and, provided the amount is small and not continuous, need not be of concern. If the discharge is whitish, this is evidence of vaginitis or metritis. The symptoms and the treatments are similar.

Vaginitis may be either viral or bacterial in origin. The visible symptom is the slimy discharge and a general discomfort you may be able to observe. It should be treated before breeding. Pessaries (large tablets inserted into the vagina) and various antibiotic injections are the common veterinary remedies.

Metritis (endometritis or septic metritis) is usually caused by staphylococcal infection and may be accompanied by a hormonal imbalance. The infection is centered in the uterus instead of the vagina. There is a whitish pus discharge similar to that of vaginitis. Metritis interferes with breeding and can interfere with the estrus cycle as well. Treatment is with pessaries inserted into the uterus (most easily done while the cow is in heat) or an iodine solution run into the uterus by catheter.

It is essential that prompt action be taken when a whitish discharge is observed. Do not wait to see if it will just go away. Advanced or chronic cases will result in sterility.

Infertility

If your cow does not come into heat after calving or does not conceive after repeated inseminations, you will need to evaluate the possible causes and take some action; otherwise, your milk supply and the usefulness of your cow will be lost. Failure to come into heat is often caused by a nutritional problem. This is especially true of the cow that has the natural tendency to put a higher proportion of her feed into her milk. My cow Belle was that way. She tended to be as thin as two boards on feed that kept the other cows in good flesh. The more I fed her, the more milk she gave, even late in lactation (feeding for more milk six months after calving is usually unrewarding). Only in the lush summer pasture did she manage to gain some weight. Belle also didn't come into heat for about five months after calving, which was a distinct nuisance. Her persistent production was something of an offset to this, however. When her nutritional level got up to where it needed to be, she put on a fine display of bulling and bred on the first insemination. But do not count on this situation. A mineral deficiency, a vitamin deficiency, or simply a poor ration will cause a cow to dispense with the reproductive cycle. This is especially a problem after winter calving. A high-yielding cow is hard put to eat enough to produce milk and restore her body condition, let alone get ready to grow another calf. Take extra trouble to feed a winter calver well and often. If appetite is a problem, try some apple cider vinegar on the dairy feed. If she doesn't like the feed very much, try to find some crushed or steamed corn in a flaked form and top-dress her feed with a scoop of it. Cows usually find this irresistible. Anything green is a big help. I've been known to sprout a pan of sunflower seeds for my cow in winter. Molasses water sprinkled on her hay will encourage a cow to eat more of it.

Ovulation may be suppressed when a cow and a calf are running together. The frequent suckling by the calf may cause prolactin levels to remain so high that estrogen is suppressed, with the result that ovulation does not occur. Separating cow and calf even half of the time may permit the heat cycle to begin. If unsuccessful, try weaning the calf.

There are other causes for infertility. The metritis mentioned above is a common one. A persistent corpus luteum on an ovary is another. A third cause is the presence of an ovarian cyst. The corpus luteum forms on each ovary after conception and stops the heat cycle for the duration of the pregnancy. But if it remains, the heat cycle does not resume after calving. Or if the corpus luteum forms after insemination, even though there is no growing fetus, there is false pregnancy: no heat periods, but no calf. A veterinarian is able to feel the corpus luteum on the ovary by reaching into the anus with a gloved hand. If your cow does not come into heat within two months after calving and you are anxious to get her bred soon, have a veterinarian examine her. The veterinarian can tell if there is enlargement or infection, or possibly retained matter (in two months' time this would have either caused illness or been resorbed, almost certainly). He can feel a cyst or a corpus luteum and "pop" them from the ovary involved. He may give a hormone injection to induce the reproductive cycle to start quickly. Your cow should then come into heat in a few days. Skip that heat and have her inseminated on the next heat.

It is important to keep a record of what you have done. If you ask a veterinarian to do an examination of your cow soon after she has been inseminated, you risk loss of the fetus. Although there must not be a corpus luteum in order for heat to occur, once pregnancy begins the corpus luteum must be left there to stop the heat cycle. If your cow came into heat normally and missed two or three heat periods after breeding, the probabilities are excellent that she "settled" (is in calf).

An ovarian cyst, in contrast to a corpus luteum, can be the result of infection, not something natural gone out of timing. The veterinarian can crush the cyst and give a hormone injection. Cysts tend to recur, causing sterility.

Failure to conceive, in cows as in all species, is due to nutritional inadequacy or imbalance more often than anything else.

Abortion

The loss of the calf before pregnancy has reached full term can be caused by an injury, by an illness such as pneumonia, by severe stress, or by brucellosis. Since you will surely have either purchased your cow from a brucellosis-free herd or had her tested, it is unlikely that this will be the cause should your cow abort. In cattle there are several other abortion-causing diseases that are familiar to dairymen and veterinarians but that do not affect humans. Nearly all these diseases are either treatable or self-limiting. Call your veterinarian and ask if you should save the fetus for tests. Then bury the remains where they cannot be found by pigs, dogs, or rodents, as one disease, leptospirosis, can be spread by these animals. A vaccination against lepto is available; you may wish to avail yourself of it if the disease is prevalent in your area.

If your cow aborts, you may start milking her. It may take some days for the milk to come in.

Bloat

It is completely normal for your cow to develop gas during fermentation of her food in the rumen (the first stomach). This gas is normally belched or passed through the digestive tract. Typically, as soon as your cow finishes eating a pan of dairy feed she will appear to lick her chops and belch, usually about three times with a loud "Haa" sound. But should this gas become trapped in foam or froth, the cow cannot belch it up. This can happen when a cow eats lush, wet green feed, especially early in the morning when the grass is heavy with dew and the air is cold or especially if there was a bit of frost. Lush plus wet plus cold – these three factors spell trouble! The resulting bloat rapidly endangers the life of a cow; the abdominal pressure stops circulation and body functions and forces toxic products into the system.

So that you can recognize the first symptom of bloat, look at the triangular area of the left flank just in front of the hip bone. Look now, while she is healthy. When your cow is hungry, this area will be depressed. After several hours of normal grazing it may be filled up to the level of the rib cage or barrel. But in bloat, this area will be inflated so much that it rises up above the level of the hip bone. The rumen is literally blown up like a balloon. Normal belching is stopped because of the frothy nature of the rumen contents. It is possible that the nerve mechanism that triggers belching is somehow bypassed. But even if the cow tries to belch, the air is trapped in minute bubbles in the froth and cannot escape. A cow in the condition described is in mortal danger. Call the vet if you are not equipped to deal with bloat, but he or she will not have a lot to offer beyond what you can do yourself. Keeping the cow standing up and walking is important. Horse people will note the parallel with colic. If you go out in the field and find your cow in this condition, try to walk her back to the house with you rather than running back alone for the treatment equipment (see below). If there is another person to help, one of you should keep leading the cow around. Try not to allow her lie down, but if she does, do everything in your power to make her get up again, including frightening her if necessary. Shout, strike her, yell.

Prepare for this emergency by having on hand a product containing poloxalene. It is available from farm supply stores and online vendors under various brand names including Therabloat and Bloat Guard. Its purpose is to reduce the froth and foam in the rumen and induce belching of gas. It is administered as a drench. If you don't have a drenching tube, the mixture can be given via a longneck beer bottle. To administer, hold your cow's nose by the nostrils. A firm grip here will immobilize the cow. Lift her head to her shoulder height but no higher or the drench may enter her lungs. Push the bottle in on the side of her mouth between the molars and cheek. Let the drench slip down her throat nice and easy. Hold her head until she has swallowed all the mixture. Unless the bloat medicine says it isn't necessary, follow up with ¼ cup of baking soda dissolved in a pint of water. This will help control acidosis, a dangerous condition. Then go back to the walking. A sound of belching will be music to your ears. You should see the rumen becoming smaller.

If belching does not quickly ensue, here is an old-fashioned trick that helps: tie a heavy rope around her head to form a halter and put it through her mouth like a bit. A USDA pamphlet from 1916 advises using a stick daubed with tar for the bit, but who has tar? The rope or stick makes the cow chomp and writhe her tongue around and, hopefully, do a lot of belching. Another excellent approach is to put a four-foot length of well-oiled or greased garden hose down her throat. Make sure it ends up in her rumen. Listen at your end for breathing. Oops, that's her lungs. When properly positioned in the rumen, the hose should emit a gush of vile-smelling rumen gas and perhaps even some foam. With the use of a funnel the hose can be used to give more drench.

In earlier editions I advised drenching with a mixture of one tablespoon turpentine in a cup of milk, but I find that nowadays few people even know what turpentine is, let alone have it handy. You can use vegetable oil instead.

If your cow lies down and refuses to get up for treatment, or if the treatment does no good, more serious measures are required, and quickly. If the vet isn't there, you'll have to decide whether to take matters in your own hands. A good investment is a device called a trocar and cannula. It is used to make a hole in the rumen and hold it open. The device has two parts, one fitting inside the other. The inner trocar punches the hole and then is withdrawn, leaving the cannula holding the hole open. You must visualize the center of the triangular area in front of the hip bone (now so distended that you must use your imagination), and stab the trocar in there. Don't be timid. If you don't have a trocar and cannula, or if the bloat is too dense and frothy to emerge through the small hole made by the trocar, you can make a hole with a knife. Use a long, narrow, strong blade such as a boning knife. Don't fool around carving away at layers with your knife. Zero in on the right place and stab. Cut a hole big enough to let the trouble escape. It will be a great mass of froth. Reach into the rumen and pull out the mess. If the vet still is not on the scene, somebody should call him, because he will be needed for the repair work. If it is a knife job, there will be contamination of the peritoneal cavity. The rumen will need to be stitched, and then the abdominal wall, and antibiotics given to control infection. But for all this, a cow saved from bloat in this manner usually heals quickly. Better patched than dead.

Bloat is mysterious. I lost a wonderful Jersey to bloat one morning when the herd was grazing in a pasture of alfalfa (always problematic eaten fresh) where they had grazed for more than a week without trouble. I was even feeding an expensive product that was guaranteed to prevent bloat. The only thing that was different about this morning, I realized when I stopped to think about it, was the breeze, which made the air perhaps ten degrees cooler. Clover and alfalfa (legumes) are always more dangerous in terms of bloat, and red clover more than white clover. When the grass is tender and rapidly grown, when it is wet or dewy, when the air is cold, and especially when the rumen is empty, you should anticipate the possibility of bloat. While you may lose some of the growth, I think it is best to get on the spring pasture early, before it can grow to dangerous length. It is grass so tender and sweet that your cow gobbles mouthful after mouthful, which can cause trouble. Mature long grass isn't dangerous. If you feed hay before letting the cow out in the morning, she will not be able to eat a lot of wet grass before stopping to chew her cud. By the time she's done, the grass should be dry and the edge should be off her morning appetite.

Overeating Disease or Enterotoxemia

Overeating of grain or other high-density feed, or feed that has been too finely ground, may lead to another form of bloat, known as overeating disease or enterotoxemia. It results when a cow overconsumes a high-carbohydrate, rapidly fermentable feed, such as milled grain. Veterinary advice directed at commercial settings usually starts out by saying "consider immediate slaughter." In other words, the prognosis is poor and the treatment uncertain, so act now while you still have something to salvage. This advice is unlikely to appeal to family cow owners. If overeating occurs with a family cow, she will most likely have gotten into the grain and gorged on sweet feed or, worse, chicken feed. It is hard to predict her odds. A large cow accustomed to generous grain feeding may survive eating forty pounds, while another cow less adapted to grain may not survive twenty pounds.

The damage follows from an overgrowth of lactobacilli, which results in acidification of the rumen. Other rumen microorganisms do not thrive in an acid environment, and the lactobacilli quickly disrupt normal digestion. Inward osmotic water flow from the blood and organs results in dangerous dehydration of body tissues. Rumen pressure increases, which inhibits gut motility; contractions slow or cease, so the grain mass stays in the rumen and provides an opportunity for a secondary *Clostridium perfringens* bacterial outgrowth, which may in turn produce a neurotoxin that further inhibits gut motility.

If you catch it early, treatment is possible. Most authorities emphasize withholding water for twelve to twenty-four hours, although one authority (*The Merck Veterinary Manual*) adds that this has not been proven. If there is evidence of bloat, put a four-foot greased hose down the cow's esophagus (the method described above) to permit evacuation of gas. The hose can also be used to administer a drench of poloxalene (again, see above) to reduce foam. Alternatively, or in addition to poloxalene, give magnesium hydroxide, also known as milk of magnesia (500 grams per 450 kilograms of body weight), in water or sodium bicarbonate (¼ cup in a pint of water). These alkaline substances will counter acidosis. Rumen pH must not fall below 5 nor alkalinity rise above 8 or permanent damage will ensue. There is the possibility of an overgrowth of clostridium species with resulting toxins; give activated charcoal prophylactically.

Most veterinarians give mineral oil or other oil both to reduce foaming and to encourage passage of rumen contents. Use the trocar only as a last resort; this form of bloat should respond to the suggested treatments. However, opening of the rumen and removal of its contents is sometimes done.

It is essential that after the cow is eating again (long-stemmed grass hay only, no grain) that rumen microorganisms be repopulated. The old-fashioned way, and it works, is to steal a cud from another animal. Failing that, probiotics should be given daily until you are sure that digestion is normal, judging from the cow patties. Relapse after days or even weeks is not unusual due to rumen or liver abscesses.

You Think It Can't Happen to You

If you feed grain, establish a security plan that cows can't breach. Try to have two doors between animals and their feed storage. Consider automatic door-closing devices. Don't settle for flimsy doors. Keep grain in locking containers. Be inventive. An old chest freezer is good for grain storage. Be sure all family members are aware of the gravity of your grain-room rules. Follow visitors around, and be a bore about repeating admonitions.

Worms

Intestinal worms are not generally much of a problem in adult cows. If your cow has a "wasting" appearance, it is possible that she has a worm infestation. You can buy an inexpensive device called a balling gun to put deworming pills down her throat. Always aim for the top of the throat, as the gullet lies above the windpipe. Mature cattle seem to develop a natural immunity after a year or so on their own pasture. This is not the case with calves, so if they must be closely confined on pasture that has been intensively used, consider worming them.

Apple cider vinegar has many advocates for both prevention and treatment of parasites. Veterinarian William G. Winter, writing for the *Stockman Grass Farmer* (April 2012), uses one gallon of cider vinegar in a hundred-gallon stock tank on an ongoing basis and reports good control of stomach worms, coccidia, and possibly even "barberpole" worms (*Haemonchus* spp.). I remain concerned about the use of cider vinegar in a galvanized tank because of the risk of leaching zinc. I would use an old porcelain bathtub or stick with plastic.

To help prevent infestations, avoid using a rotating or flail mower on pasture, as this puts manure all over the place. Undisturbed, a lush ring of grass grows around each cowpat, and this will be avoided by cattle for a year or longer, breaking the cycle of reinfestation. (See chapter 13 for more on pasture management.) Note also that composting the manure destroys most parasites.

Lice

Lice sometimes infect cattle when they are tied up a lot or conditions are damp. Lice prefer the area around the tail, head, and ears and in the folds above the udder. I have had good success in treating cattle lice using diatomaceous earth. This is sold with garden supplies and is a nontoxic powder similar to ground limestone. Just shake it on and rub it in. Sprinkle it all along the spine and get it up into the folds.

Sore or Cut Teats

Cuts, scratches, and abrasions on teats are most commonly inflicted by fences or junk left spread about. This can be prevented by keeping the fences firm and the areas where your cow lives and grazes clear of rubbish. A cow sometimes steps on her own teat while arising and inflicts a severe cut. Minor injuries can be treated effectively in the same way you would treat them in humans. The self-adhesive bandage called vet wrap works well. There is nothing better (for man or beast) than vitamin E for promoting rapid scar-free healing of a wound. Coat the wound with wheat germ oil, or prick vitamin E capsules and squeeze the oil directly on the injury.

The various udder creams sold in feed and farm stores are useful when teats become dry or chapped, as often happens in winter or when they are left wet.

If your cow should receive a serious teat or udder wound, call for veterinary aid quickly. I have seen some remarkable repairs accomplished successfully when the work was done right away. Apply the vitamin E while awaiting the vet.

A lactating cow with an injured teat must still be milked. A severely cut teat or one that has been sutured cannot be grasped for milking without the risk of breaking open the wound. In this case you can insert a small disposable plastic cannula into the teat orifice at milking time. The milk will then drain out during letdown. After milking, pull out and discard the drain. The disposable cannulas are sold in bubble packs by the hundred, or you can get them from your vet.

If the cut was not serious enough to require suturing, you may be able to milk without the use of a cannula. Once letdown is taking place just grasp the teat firmly and begin milking. After initial flinching a cow often seems not to be much bothered.

Sometimes a cow will have sore teats for no apparent reason and the problem will continue and often get worse despite continued applications of udder cream. The problem may be udder impetigo or cowpox. Impetigo is a staph infection. Look for small yellowish pimples at the center of the sore spot, which is usually at the base of the teat. These typically enlarge and scab over, then burst and ooze blood and a clear fluid. The infection is rapidly spread by your hands when milking. Udder creams are totally useless against it. Before I found the effective treatment, I tried every kind of udder ointment on the market with no success. Instead, treat impetigo with a triple-dye solution, such as Dr. Naylor's "Blu-Kote." The trouble will be gone in a few days. Iodine is also somewhat effective.

Cowpox is less common than impetigo. The pimples are red rather than yellow, and there is more fluid when they break, leaving a ring-shaped sore. Fortunately, you don't need to be certain which it is, cowpox or impetigo, because the treatment is the same for both.

Bovine Spongiform Encephalitis or Mad Cow Disease

Bovine spongiform encephalitis (BSE) is one of a family of brain diseases that are slow to develop, debilitating, and invariably fatal. Scientists continue to disagree as to the nature of the disease agent. Extremely small rod-shaped structures called prions are found in diseased brains, but it is not known whether they cause the disease or are an artifact. Some researchers believe a virus is involved, though they have yet to find one. The mode of transmission is by no means fully understood, but it is well agreed that in cattle most cases result from the practice of augmenting their feed with recycled slaughterhouse waste, dried and granulated. This waste includes the remains of many species besides cattle.

Transmissible spongiform encephalitis (TSE) is the term used for the entire family of these diseases, which include kuru, Creutzfeldt-Jakob disease (CJD), scrapie, transmissible mink encephalopathy (TME), and BSE. The similarities among these diseases are greater than the differences; all result in the characteristic spongy-looking brain found at autopsy.

Kuru was identified among New Guinea highlanders who followed a ritual cannibalistic practice of eating the brains of the dead. The etiology of kuru was first recognized by Carleton Gajdusek following many years of study. He was awarded the Nobel Prize for his work in 1976.

CJD appears to be idiopathic; it arises spontaneously in all parts of the world, causing about one in a million deaths.

Scrapie has been known for centuries. It is found in sheep and goats and probably some other ruminants such as deer. The mode of transmission is not known. In Britain, where it is endemic, it is generally believed not to infect humans despite having been transmitted experimentally to monkeys.

Mink encephalopathy (TME) has killed many mink in the United States but so far has not been shown to migrate to other species. This may be due only to good luck, since experimental infection of cattle with TME has been fully successful.

The work by Gajdusek on kuru and later on CJD combined with work by British researchers on scrapie has formed the basis for our understanding of BSE, even though our knowledge of these diseases remains incomplete. BSE was first identified in Britain in 1985. Studies of the disease were uncoordinated, and management by the British government emphasized soothing pronouncements and protection of investment over prevention. BSE has now been found in several European countries. British beef quickly became suspected of causing the outbreak, although it is not clear that BSE is transmitted via ordinary beef or dairy products. It is unquestionably transmitted by brain and other neural tissue. These tissues were banned by the British government from inclusion in meat waste intended for use in feed. However, compensation to those financially injured by loss of markets (farmers, slaughterhouse owners, and exporters) was slow, erratic, or nonexistent, and consequently compliance was slow, erratic, or nonexistent.

Even where compliance is attempted, slaughterhouse practices are often very messy. It would be difficult to be sure that no brain tissue contaminated beef carcasses, let alone that it was not included in offal destined for recycling into bovine, swine, chicken, or pet food.

Diseased brain tissue remains capable of causing infection even after sterilization at very high temperatures and long soaking in formaldehyde, and it survives irradiation. Although BSE (the bovine form of the disease) has not been reported in the United States, one woman in Florida was infected with CJD (Creutzfeldt-Jakob disease, the human form) after receiving a corneal transplant from a victim of that form of TSE.

Importation of British beef to the United States is now restricted, as is importation of all live ruminants including zoo animals. As an initial step in protecting the public from the largely hypothetical spread of BSE in U.S. beef, the FDA has restricted recycling of ruminant remains (cattle, sheep, and goats). However, laboratory transmission has been accomplished between many nonruminant species, including primates and all the usual rodents. Direct injection of brain tissue into living brain results in disease in most recipients. Infection from feeding brain tissue is inconsistent but clearly occurs.

Some calves in Britain have been found to be infected, but it is not known whether infection was acquired from their mothers or from ingesting a protein supplement that is sometimes fed to calves. It appears that infection with BSE can occur both horizontally (to herdmates) and vertically (intragenerationally) to offspring. Pursuing the truth about BSE transmission in cattle is slow work because the disease develops slowly in its victims, often not appearing for seven or more years in humans. It appears to take nearly as long to develop in cattle unless injected directly into the brain. This is why it has been seen in dairy cattle rather than beef cattle; many dairy cattle stay in the herd until they are ten years old. Beef cattle are killed at sixteen months to two years.

Brain tissue of laboratory animals deliberately infected with BSE is capable of causing disease in others long before florid symptoms occur in the host. If this is also true of cattle, swine, and sheep, then opportunity for the spread of BSE through the recycling of apparently nondiseased animals and animal parts clearly exists. The disease exists in every country that follows the practice of augmenting the diet of farm animals with animal protein, which is to say every country with a meat industry. Currently, this animal protein product includes dead animals euthanized at animal shelters.

Cats are known to develop TSE. Pet food contains animal by-products as a protein source. Poultry feed contains a great deal of animal protein supplement. Although birds appear not to contract the disease, it is thought to pass through their gut unaltered; then chicken manure is spread on row crops. Chicken manure is also dried and included in various animal feeds, including dairy feed and pet food, under the euphemism "poultry digest."

With this information in hand, Carleton Gajdusek began to advise against the use of bonemeal in gardens. It is extremely dusty. What goes into your nose goes straight to the brain via the olfactory nerve.

Gajdusek and a number of other researchers came to believe that the disease agent in TSE is chemical in nature, which would account for its immunity from the usual sanitizing agents and autoclaving. But how does a chemical agent, a mere group of lifeless molecules, grow and increase, as the disease agent for TSE clearly does? Researchers suspect it does so by inducing a cascade of molecular reactions not unlike that described in Kurt Vonnegut's possibly prescient 1963 novel *Cat's Cradle*, in which water is seeded with a variant crystal called ice-nine, thus raising the freezing point with deadly effect. A variant form of an ordinarily molecule may cause a dramatic change in an otherwise stable solution. Something very similar occurs in a common high-school lab experiment through a process called seeding; a supersaturated solution of sodium thiosulphate (photographic hypo) remains stable until "reminded" of how to crystallize by the addition of a few pure crystals. The entire solution then immediately crystallizes. Gajdusek showed that the amyloidal plaque characteristic of several brain diseases, including TSE and Alzheimer's, is a crystalline protein. Many vital normal physiological components also have a crystalline structure, including cholesterol.

For the foregoing review of current knowledge of TSEs, including BSE, I have relied principally on *Deadly Feasts* by Richard Rhodes (Simon & Schuster, 1997) and "Pathological Science," an essay in *New Yorker* magazine, December 1, 1997, by the same author. Rhodes provides balanced reporting based on the facts he has assembled, and he digs hard. He identifies his sources. Knowledge of BSE remains incomplete and will continue to be politically charged due if nothing else to the money and careers impacted by this insidious disease.

FOOD SAFETY

To protect us against food-borne illness, government agencies in the United States have relied primarily on instructing consumers to wash their hands, cook chicken and meats to a high internal temperature, and refrigerate food; in other words, make it our job to avoid getting sick from contaminated food. Little has been done to force changes in food processing. The dairy industry is something of an exception to this rule. Due to its structure, within which independent unorganized dairy farmers operate from a position of weakness, the public has had the benefit of many decades of mandated culling of cows with the old-fashioned diseases, besides the added safety of pasteurization. In the case of BSE, now for the first time only profound changes in the food industry itself are capable of ensuring public safety.

Nothing you or I can do in the way of cooking or washing will protect us from BSE. Going vegetarian is a short-sighted response; eliminating proper food and then relying on the latest supplement to make up the deficit is inefficient, costly, and ultimately not effective protection against either transmissible diseases such as the flu or degenerative diseases such as cancer. This is clearly understood in animal nutrition. Indeed, the whole point about feeding animal-derived protein supplement is that animal protein, whatever its source, works better than anything else to support the goals of growth, reproduction, proper development, and resistance to all types of disease.

Compelling farmers to augment feed with soy instead of meat protein is an option that has been suggested, although there are drawbacks. Soy is more expensive than a meat-based supplement and is only about half as potent a protein source. It has limitations as a support for growth and reproduction. For example, in rearing baby calves, where growth is critically important and outcomes readily apparent, soy milk emerges as a less effective "baby formula" compared to replacers made with real milk, though the soy replacers are cheaper than milk-based replacers. Meat-based replacers are also available, and they are said to be the equal of milk and less costly than milk.

What You Can Do

Here is what you *can* do. Buy a cow from a herd that is not using animal protein supplements. If you have the space, buy some beef critters too. By buying your calves from a grass-fed or certified organic herd, you can be sure neither they nor their mothers have had access to meat supplements. You will then be in a position to supply completely safe meat to your own family and perhaps a group of grateful friends.

In my view, only the reestablishment of local animal husbandry and local abattoirs can provide us again with safe, affordable food.

CHAPTER 17

Safety around Your Cow

Working with any animal, possibly excepting guinea pigs, contains the potential for injury. I have never been seriously injured by a cow but have come close. As in most endeavors, injuries tend to occur when you overlook precautions or takes shortcuts. In working with animals, injuries may also occur when behavior typical of the species is either ignored or not understood, as when an ordinarily well-behaved dog bites a toddler who puts her fingers in his food. Here are some observations about cow behavior and some hazardous situations you should avoid.

The very first rule of defensive cow management is to wear good shoes. A cow doesn't even seem to know when she's on your foot, and your imprecations will just make her gaze gently around with a "How's that again?" expression.

Cows are supreme creatures of habit and will meet any changes with obstinacy. When making routine changes, plan ahead. Shut any gates or doors you do not want the cow to go through. Never stand in a gate or passageway through which she is passing. A cow contains a lot of mass, and once it is moving forward it is likely to keep moving. She is wide, and she can't suck in her waistline. Stay out of her direct path.

Put her food out *before* you let her in and *close* the grain storage. If she sees her grain bucket in your hand she is likely to knock it away and you won't be able to stop her.

Leading

You can train a calf to lead (see page 103 for a discussion of this training). But a cow purchased from a dairy herd probably will not understand being led. She will have been habituated to walking soberly to her own stall and sticking her head in her own stanchion. Teaching a mature animal to lead is not easy. But once she is fond of you and gets the hang of what you want, she will cooperate to some extent. To lead her with any hope of success you will need a cow halter. This has straps that cross under the chin and a strap behind her ears with a throat latch. There is a ring under her chin for a snap where the straps cross. A horse halter cannot be used. She can wear the halter all the time if you like. To lead a reluctant animal, a simple collar will not do; there is no way you can get enough leverage on her head. A cow follows her head; any way you can turn her head, that way she will go. But her neck is short and strong, so your leverage must be as far forward as possible, right under her chin. Trying to lead a reluctant cow is a situation to be avoided. Put up good fences, keep gates closed, and plan to lead her with a bucket of grain if you have to.

On the other hand, cows are easily driven (unlike goats!). Get behind her with a long stick, try not to get her running, and quietly move the stick from side to side to let her know which way you *do not* want her to go. She will move ahead of you and away from the stick.

If a cow is dragging a rope, don't expect to control her with it unless you are able to snub it with a double wrap around something very strong such as a tree. A cow, because of her shorter neck and legs, is actually stronger than a horse of the same weight. That is why oxen have been used for plowing and hauling for many centuries. Keep your fingers out of the way of a running rope and *never* wrap it around your wrist. Just let her go.

If you must attempt to lead a confused or frightened but otherwise domesticated cow without help, here is what you can do: The lead rope must be short, no more than five feet long, and it must be on a clip or running loop under her chin, not around her neck. A knot at the far end of the rope is helpful. Take hold of the rope under her chin and just far enough back so that your elbow gets her in front of her shoulder. Pass the free end of the rope across in front of you to your other hand. Walk as close to her neck as possible. She can't hurt you in this position. She cannot rise up or strike as a horse might. If she walks too fast, jab your elbow into her shoulder and try to keep her nose down just a little. If she turns away from you, give a good jerk on the rope. If she turns toward you, maintain your handholds and back around in a circle until she is again pointed forward. Using this method I have led home a cow that jumped the fence while in heat. There were woods and a ditch on one side, a road with logging trucks on the other. I was not hurt and neither was the cow, but you may be sure I improved my fencing. Furthermore, I kept good track of her next heat and did not let her out of her stall during it. Prevention is best.

If you are familiar with horses you will recognize the similarity of this method of leading to that used with a horse. The difference is that with a frisky horse you aim to keep the head down and neck arched so he does not rise, strike, or bite. A cow never rises or strikes and cannot bite, but if balked she will kick up her heels and twist. So don't pull her head down too far or you will inhibit forward motion and induce her to kick up her heels. The first impulse of a frightened or confused cow is, like a horse, to run away. So you are permitting enough forward motion to keep the cow's mind off balking or struggling, and you are pulling the nose down sufficiently to prevent the stretched neck and raised head of the racing trot.

Another distinction between a horse and a cow is that a fleeing horse will not enter the woods if there is any other choice at all. A cow will dive right in (she isn't called *boss* for nothing).

Birth

Each year two or three farm wives are killed by interfering between a cow and her newborn calf. I have seldom had an otherwise friendly cow turn aggressive when she has given birth in her familiar stall. But I don't turn my back on her. If she gives birth down back in a field near the woods, watch out. Maternal instinct is jump-started not solely by the act of giving birth but also by the surroundings, the environment of evolutionary addictedness. If a cow has her calf all by herself out in her natural setting, she figures it is her responsibility to defend it. *You* figure the sweet little thing will get lost in the woods and you had better carry it home. Don't. Go get some help. She will still be there when you get back. One of you will have to carry the calf, because in these circumstances no earthly force will drive the cow from her calf and so one of you must carry the calf to lure her along. The other of you must watch the cow to make sure she does not knock down the calf abductor. It is tricky, because a cow seems unable to see the calf in your arms, so you have to walk in such a way that she keeps in nose contact with it. If this connection is lost, she will run back to the spot where she gave birth and will not be moved from that place until you carry the calf all the way back and start over. If the cow leads well and the calf is on its feet, you can probably lead her along and the calf will follow. She will be satisfied with this arrangement as it is what calves are meant to do and she can see it out of the corner of her eye.

I was once compelled to carry home a calf by myself because the cow had given birth on a little island of field cut off by a rising river. One false move by the calf and it would have drowned. In that case I hurried along without regard for the cow, who predictably remained at the birth site. I then returned for the cow, with help to drive her, by which time she was standing in the water. It is usually a first-calf heifer who gets into these situations. An old, friendly cow often seems to welcome your help. But don't take a chance. Once an angry cow has knocked a person down, even a hornless cow knows how to grind her head on her victim's chest, with sometimes fatal results.

Heat

Another situation to watch out for is the cow in heat. During the early stages of heat, the cow has an impulse to mount other cows. If you don't have any other cows, guess who she will be interested in! Sometimes the only way you know an "only" cow is in heat is when you see her standing with fixed attention, waiting for you to turn your back. An ardent cow is not going to go into attack mode in case you stumble, so any injuries will be caused by what you stumble over. A mounting cow puts her head up high and leads off with her chest, so even if she has horns the chief damage will probably be to your dignity, but don't say I didn't warn you. Once the first clue I had that my heifer was in heat was when a set of front hooves appeared on my shoulders.

Kicking

The propensity of certain cows to kick over the milk bucket is legendary. The only worse thing is when both the bucket and you catch the blow. The kick itself is not life-threatening but can knock you off your stool. She can also lift her foot up and bring it down on the edge of the bucket, which could badly hurt your hand if it's in the way. The edge of her hoof is fairly sharp and can inflict a cut if brought down on a hand on a hard surface. Some cows are sufficiently athletic that they can kick back with the front foot. This is a weaker kick but just as annoying.

Here again, prevention is the best course. Most cows have a fine dairy temperament, are cooperative, and grow to love you. Usually there is a good reason for kicking. A cow, particularly a heifer, may kick quite a lot when her bag is painfully swollen with milk after calving. Sometimes the first you know of a teat injury is when you grab it and she kicks, and cows kick very fast. I'm fast, too, and have not been kicked in years. But that is mostly because I've learned to sense the changes in posture and attitude that precede a kick. If I have a cow that is inclined to kick, I take immediate steps to forestall it. Cows like to do the same thing each day. If she got in a good kick yesterday, she may do it today just because that is what she does. If she did not kick today she probably will not tomorrow, unless something worries her. The things she hates most are an unfamiliar milker, shrill running children, and dogs. Introduce any new milker tactfully and train the children to be as quiet as possible and she will accept them. But her fear of dogs is hardwired, and a cow will seldom feel relaxed when any dog is around, especially while she is tied up and defenseless. You may have to keep the dog out of the barn. A cow also does not like a cat swarming around her legs.

If tact fails to prevent kicking, get one of the appliances designed to prevent kicking. You'll find several types available at any large animal supply house. You have a right not to be kicked. (Kicking is discussed further in chapter 3, "Milking Your Cow.")

Apart from new motherhood and heat, cows are mild-tempered and focused on their own internal bliss. You quickly come to understand why they are revered in India as partaking of the divine.

A Cow Does Not Bite

I am occasionally asked whether a cow bites. A cow, like all ruminants, lacks upper incisors and it certainly doesn't have canine teeth. Artists hired to decorate milk cartons or do other cow illustrations often seem ignorant of this basic physiology and depict a cow with upper teeth like a horse. So I guess the public must be forgiven for thinking a cow can bite. A cow does have molars, so if you are giving medication you need to watch out for the back teeth. And those lower incisors on a newborn calf, unworn by grazing, sometimes cut the teat.

Mobbing

Mobbing refers to an activity practiced by many species whereby a large number of otherwise defenseless creatures move as a mob to discourage predators or intruders. You can often see little birds driving away crows this way. Sometimes a large group of cattle will do this. Walk purposefully away. Running is to be avoided if possible.

Bulls

Just do not ever get anywhere near a dairy bull. They can never be trusted.

CHAPTER 18

Cattle Breeds

Most of the common cattle breeds in the United States are European in origin, with the majority originating in the British Isles, which for centuries led the world in animal breeding.

Dairy Breeds

Although Britain has provided the majority of breeds, Holland gave us the Holstein-Friesian, the black-and-white breed that now accounts for most of the dairy cows in the United States.

Holstein-Friesian

These Dutch cattle, now the major dairy breed not only in the United States but in virtually all Western countries, were first brought to the United States in 1795. They have been bred here and in Canada to produce a volume of milk unmatched by any other breed. They are large, splotchy, black-and-white cows that also make quite creditable beef. The cows are often crossed with beef breeds, especially Hereford, and the calves are then raised for beef, providing the dairy farmer with an additional cash crop. Such calves are black with a white face.

Brown Swiss

One of the oldest breeds in existence, often valued as dual purpose because of the cattle's large size. Brought to the United States from Switzerland in 1869. Their markings are virtually indistinguishable from those of the Jersey.

Ayrshire

Cherry red to brown with white in color. Their color patches are smaller and more broken than those of the Holstein-Friesian, and they are slightly smaller. If allowed to grow, their horns have an elegant lyre-shaped form. The carcass is of high quality and milk production is high; they may be considered dual-purpose cattle. Brought to the United States in the late nineteenth century from Ayrshire in southwestern Scotland.

Guernsey

The cow closest to the Jersey in the butterfat content of its milk, and in origin, coming from the island of Guernsey in the English Channel, about twenty-five miles from Jersey. Brought to the United States starting in 1830. Larger and generally heavier-boned and more rangy than the Jersey. The cream and butter have a very beautiful color. Back when there was more choice in quality of milk in the United States, the Guernsey dairymen sold their product under the name Golden Guernsey.

Jersey

Within sight of Normandy on the coast of France in the English Channel lies a group of four small islands – Jersey, Guernsey, Alderney, and Sark. The islands are the native home of two distinct breeds of dairy cattle: the Jersey and the Guernsey. The island of Jersey, the largest of the group, has an extremely mild climate, and its cattle can be outdoors most of the year. On both Jersey and Guernsey, the eponymous breeds are the only cattle allowed by law on their respective islands to this day.

The Jersey breed probably has Eastern origins, by way of France, for the cattle are similar to those found in Asia. There is the same whitish ring around the muzzle, coloring of the coat, richness of the milk, and ability to acclimatize to extremes of heat and cold. Some of the Asian cattle have humps and do not give much milk, yet their progeny by Jersey bulls lose the hump, look like Jerseys, and milk much better than their dams. Another possible indication of the Eastern origin of the Jersey is that pictures of cattle found in tombs in ancient Egypt show a strong likeness to the breed.

Why I Choose the Jersey

The virtues of the Jersey that make her an ideal family cow are as follows.

Milk Quality

She has the highest level of milk solids (protein and minerals) and vitamins, as well as butterfat, of any common breed of cattle.

PERSISTENCE OF PRODUCTION

She is a persistent milker, with the tendency to produce at a more even level throughout her lactation than is the case with most breeds. This is an especially valuable trait in a cow that is the only one providing for the family needs.

LONG MILKING LIFE

The Jersey starts in milking at a younger age than other breeds and can be expected to continue for a dozen years or more. In a study of more than a million cows of the five major breeds – Holstein, Brown Swiss, Ayrshire, Guernsey, and Jersey – Jerseys had the longest productive life. There are accounts of Jerseys successfully calving at age eighteen.

ADAPTABILITY

She is adaptable to a wide range of climatic conditions and is naturally strong in constitution. While smaller and finer boned, she is usually tougher in a stress situation than a Holstein.

HIGHEST RETURN FROM BOTH LAND AND FEED

With her small size, the Jersey gives the highest return per acre of land and is marginally superior to other breeds in return on feed consumed. In other words, you get more milk per pound of feed. Stated another way, you are feeding less cow for the volume of milk produced.

The ideal Jersey is a wedge-shaped, beautifully formed, graceful animal with plenty of depth through the heart, good body capacity, and a large udder with well-placed teats. In size the Jersey cow rarely exceeds 900 pounds, averaging 750 to 800 pounds. The characteristic marking is the mealy ring around the nose, and the nose itself is black. (Most other breeds have a pink nose.) Coat colors are various shades of fawn, mulberry, broken fawn and white, mulberry and white, and silver.

An early puberty is one of the breed's valuable assets. Heifers (young cows not yet calved for the first time; the term is also used to refer to cows that have calved once) are usually calved at age two or soon afterward. Jersey heifers are as mature at age two as those of other breeds are at two and one half years. Although the Jersey begins her working life early, this does not impair her ability to milk well and for as many (usually more) years as any other breed. A heifer will produce 6,000 to 12,000 pounds (750 to 1,500 gallons) of milk in a 305-day lactation. When a Jersey is two and a half years old (the age when heifers of other breeds are first calving) she may already have given 4,000 to 7,000 pounds of milk. There are many Jersey dairies with averages as high as 18,000 pounds of milk, which compares favorably with Holstein production.

Total farm profitability is usually better with the Holstein cow, which may produce 1,800 gallons (not pounds) of milk a year. The Jersey produces perhaps 900 gallons in a year. Simple arithmetic shows why this is so. The Holstein eats half again as much as the Jersey, takes half again as much land and housing space, but gives nearly twice as much milk by volume. Payment for milk in the commercial market is principally on the basis of volume. Although the industry has moved toward protein-based pricing, there is little reward in our fat-phobic culture for the extra cream. Production achieved by fewer cows means lower labor costs. And the Holstein calf crop is worth a great deal more because the calves are twice the size of Jersey calves and the fat is pure white. The fat of Jerseys and Guernseys is yellowish due to the presence of beta-carotene, which is absent in the Holstein. Because of the color of the fat, the Jersey carcass is disdained in the market. Jersey bull calves are often given away, while Holstein bull calves sell for $80 to $100. We have discussed milk quality, but the dairyman has to survive in a very difficult business and is forced to go for quantity. In yield per animal in relation to feed consumed, yield per animal in relation to acres grazed (feed conversion rates), length of working life, and the regularity of breeding, Jerseys come out ahead of Holsteins on all counts and thus are better suited to the household economy. But by factoring in labor costs, which are lower per gallon of milk for bigger Holsteins, and adding the greater carcass value, Holsteins remain the commercial choice.

A Jersey weighing 850 pounds might require 125 pounds of grass (wet weight) per day. On this basis, an acre of good grass supplies the feed requirements of a Jersey for 400 days or of ten Jerseys for 40 days in the year. A dairy cow of a larger breed, weighing perhaps 1,350 pounds, requires 200 pounds of grass a day, and an acre will provide its feed requirement for 250 days, or it will supply ten cows for 25 days. This variation also applies to the amounts of hay, silage, roots, kale, and other foods consumed in winter. In a family cow, this reduced level of feed requirement is a distinct advantage, brought about by the smaller amount of cow that must be maintained. Another asset of smaller size is that a Jersey takes up about three-fifths the space required by a larger breed and is easier to handle.

As noted, an important characteristic of the Jersey is her ability to keep up a steady flow of milk throughout her lactation. In many cases a cow commencing with a yield of five gallons a day is still giving three gallons thirty-nine weeks after calving.

Jersey milk has at least 20 percent more cream than Holstein milk. It also has 20 percent more minerals, protein, and vitamins. Although the milk of no other common breed exceeds the nutrient value of Jersey, the milk of all other cattle breeds is superior in nutrient content to Holstein milk, Holstein milk being tops only in water content. When Holstein milk falls below the legal minimum for solids it is fortified with added powdered milk (there is no labeling requirement for this), and there is a move afoot to require such fortification of all milk in order for it to achieve a protein standard more akin to that of Jersey milk, thus adding a huge multiple to the number of cows represented in each gallon of milk.

Cost

Several of the less common breeds offer most of the traits found in the Jersey, but they are more expensive. You will have to pay extra for their rarity, often a great deal extra, for characteristics that may not be very meaningful in the long run.

Intelligence

I recommend the Jersey for all the reasons discussed above. It also appears to me that the Jersey has the highest intelligence among cattle and the most interesting personality. People who have not known any cows always express astonishment when I declare that cows are as intelligent as dogs or horses. They thought cows were just walking meatloaf. I tell them of a cow I once had that would always come to the gate and bellow to tell us if another cow was calving. On one occasion she led us a long way into some brush to a cow calving down in a ditch. In the old days when a lot of people had cows, stories such as this and even better ones were common. Although my experience leads me to believe the Jersey to be especially intelligent and charming, cows are far from stupid.

Intelligence and charm are important in a family cow, because working with a responsive animal makes the relationship more enjoyable. In most books on dairy cattle there are tables showing comparisons of the various dairy breeds. These tables compare the milk quantity and quality, the size of the animals, the ages at first calving, and so on. There is also a column on "disposition" or "temperament." The Jersey is usually listed as "nervous." That is a superficial view. The Jersey is simply very alert and curious.

If, for reasons of availability or preference, you choose another breed, this book will be equally useful in establishing your own cow economy. For me, the quality of Jersey milk is the most persuasive factor and best answers the question of "Why a Jersey?"

Dual-Purpose and Beef Breeds

Dual-purpose refers to breeds commonly used for both meat and milk. All breeds began as dual purposes. *Naturally polled* refers to animals born without horns.

Hereford

Sometimes called the Whiteface. A stocky breed that is a rich red-brown in color, with a white face, often with curly hair around the forehead. From the county of Hereford in England. Brought to the United States early in the nineteenth century. Popular in most areas of the country as a reliable producer of high-grade beef. Occasional individuals make fine family cows.

Shorthorn

Called Durham in some areas. Red, red and white, or roan and more rangy looking than the Hereford. Formerly a dual-purpose breed, though distinctive milk and beef types have now been developed. From Durham County in northern England. Also brought to the United States early in the nineteenth century. It's the one you saw in *All Creatures Great and Small.* The Milking Shorthorn branch of the family has become rare. The milk is of very high quality.

Aberdeen-Angus

Also known as the Black Angus. Black, naturally polled cattle brought to the United States in 1873 from Aberdeenshire, Scotland. They are purely a beef breed, hardy, slow growing, and not easily handled, but with superior beef quality. Cows from this cross are pretty good milkers. As such, crosses with the Holstein are very common as a source of dairy beef.

Devon

Sometimes called Ruby Reds, and they are, indeed, ruby red in color. Similar in shape and size to the Angus. Brought to the United States one hundred years ago from Devon, in the West Country of England. Historically were dual-purpose but now have been bred divergently for milk or beef. Considered fine draft animals.

South Devon

A very large dual-purpose breed, weighing on average two hundred pounds more than a Holstein, and a creditable producer of high-quality milk and meat of outstanding flavor. Red-brown in color. A recent import to the United States, originating in the southern portion of County Devon, England. They have an interesting prehistoric appearance.

Dexter

An Irish Devon-Kerry cross. Short-legged, small, dual-purpose breed. The Dexter is rapidly gaining in popularity for its small size and creamy milk. The cream is a little slower to rise, making for good table quality.

Highland

Spreading horns; long, shaggy, tawny red coat; and a thick mane. From northern Scotland and used to heavy weather. Slower growing, but able to make it on rough grazing. Marvelous to look at!

Longhorn

There are both British and American versions. The British Longhorns have mostly disappeared into the Shorthorns, while the American variety is descended from cattle brought from Spain and is becoming rare, even though it is associated with the lore of Texas and the Wild West. Although the Longhorn was able to successfully become feral and increase in the American Southwest, it remains a distinct breed (like Dobermans or cocker spaniels among dogs) and is not a cattle ancestor, as for instance the wolf is to the dog.

Red Poll

Medium-sized, naturally polled dual-purpose cattle, blended from indigenous cattle of Norfolk and Suffolk, in eastern England.

Charolais

Large, white or cream-colored cattle from France. Fast-growing animals that have been imported into many countries in recent years. Popular in the United States for crossing with other breeds.

Brahman, Brangus, Santa Gertrudis

The hump-shouldered Brahmans are from India (in some countries they are known as Zebu). Brangus is a cross of Brahman and Angus. The Santa Gertrudis was developed in Texas by crossing Brahman and Shorthorn, with a view to obtaining an animal better able to stand the heat.

Beefalo

A fertile cross of the buffalo with cattle has been achieved. The virtues of this big animal are its abilities to grow rapidly on rough forage and to take the living in rough conditions. We raised a Jersey-beefalo cross and then butchered her at one year of age because she was a troublemaker. The meat had a mature flavor not ordinarily present before eighteen months of age and was of excellent quality. However, claims for nutritional superiority of beefalo or pure buffalo meat based on the absence of marbling are disingenuous. The same claim can be made with equal validity for meat from any dairy breed because only beef breeds produce marbled meat. Dairy beef will be available at a fraction of the cost. Always be suspicious of any claim that an animal product is "lower in cholesterol." It is always code for lack of marbling or smaller serving size.

CHAPTER 19

Your Cow Economy

Does it pay to keep a cow? For the last fifteen thousand years and more the answer was too obvious to bother asking. Cattle, more than anything else, were synonymous with wealth. Is the world so different now? We certainly do our cost accounting differently today. Once you have your own cow you will modify the following numbers to fit your circumstances. This chapter can get you started. In the following example I am assuming 165 days of grazing and 200 days of hay feeding.

Additional costs for trucking, fencing, housing, veterinary expenses, pitchforks, and so on are important to keep track of, but you will get a clearer sense of your cow economy if you log them separately.

If the cow is on a no-grain regimen you can deduct the cost of the grain, but you will then need to deduct 20 percent of the milk production.

If the cow freshens at five gallons a day, you dry her off after 290 days, at which point she was giving two gallons a day, that's an average of about 3.5 gallons per day or 1,015 gallons per year. Cows in commercial herds do a lot better than this and yours may too. But at $4 a gallon that's $4,060 worth of milk. Deducting the costs of the cow and her feed, you are, in theory, $1,035 ahead.

ITEM	COST
Cow	$1,000
Hay, 300 bales @ $5 each	$1,500
Grain, 1,200 pounds/year	$372
Minerals	$50
Loose salt	$50
Kelp	$53
TOTAL FOR YEAR	**$3,025**

In the second year, when the cow is already paid for, your costs are only $2,025. If you have a $600 heifer calf to sell, you may consider yourself $2,635 ahead.

If you are raising a steer you won't want to butcher before eighteen months. Both butchering costs and what he may bring at auction vary greatly, but you ought to be able to count on 450 pounds of meat if you choose to butcher.

If a family of four uses a gallon of milk a day, there is an average of two and a half gallons a day to sell, make into value-added products, or feed to other livestock. At this point, if you consider value-added products, the cost accounting can get quite interesting.

If raw milk sales are legal in your area, the two and a half gallons can be sold at the farm gate for at least $6 a gallon. Or you can skim the cream and sell it either as cream or as butter, making from $10 to $20. You can make cheese from either skim or whole milk. Skim milk or excess whole milk can be used as a significant part of the diet of chickens, pigs, or calves, or as fertilizer. Clearly butter is what you do with cream that doesn't sell. But I would never sell much butter. It is too valuable in the family diet. I like to know what my dairy products are worth, but I keep a cow so that we can have all the high-quality dairy products we want.

Don't forget that the cow's manure is also valuable, either as fertilizer on your fields or to sell. Where I live, dried cow manure sells for $7 for a twenty-five-pound bag. But as with butter, I consider the manure to be too valuable to sell.

Note: Calves need whole milk at least until they are grazing well. Many people wean the calf at four months. Other folks prefer to keep the calf on milk for the freedom of once-a-day milking.

My friend who makes cheese every day has two cows. They are on pasture alone and receive no grain, and he milks only once a day and dries them off in winter. In winter they get only hay. He gets about five gallons a day from the two of them in summer and is able to make this into a cheese weighing about two and a half pounds. He sells these for $18 each. These cheeses represent pure profit. We are not counting labor in these examples. He owns the fields, which would be idle and unproductive if he did not have cows. With different management (twice-daily milking, augmented feed) he could of course be making twice as many cheeses.

As explained in chapter 9, "Feeding Your Cow," under some circumstances it is feasible to keep a cow on grazing and hay with no grain. And it is also well within the scope of a small family to grow a half acre of corn or small grain to provide energy for a cow. Purchased inputs can be greatly lowered with either of these feeding modes, although on most grazing programs production goes down pretty sharply.

These examples suggest ways of rationalizing the income and outgo from a cow. Over the years I have received wonderful reports from cow owners in amazingly varied circumstances detailing their experiences. It is hard to lose money on a cow. But these examples do not begin to encompass the possibilities inherent in a dairy cow. If you sell nothing, not even the calf, you can easily use up her full production feeding pigs and chickens. Pigs and chickens can be fed at almost no expense if you have some land they can run on and they have access to cow manure; they will turn it over to find undigested grain or bugs already at work. Weeds from the garden, fall apples, bread crusts, and much other free food will do as basic diet for pigs and chickens and will keep them going. Muscle growth, egg laying, and reproduction are dependent on a source of animal protein. This your cow can provide. Buttermilk, skim milk, and whey in their diet make success possible with pigs and chickens at virtually no cost. You'll get pigs with both muscle and fat, and sows capable of large litters, should you wish to breed your own. You'll get hens that actually lay, not just take up space on the perch, and cockerels that make meaty dinners. With access to summer cow pasture, both chickens and pigs will get a significant part of their protein from bugs.

Ruminants make milk and meat on a diet of plant products. On a similar diet, other animals only fatten and make poor growth; they need a true protein source such as milk before they can build muscle and reproduce. The cow is thus an engine capable of driving the entire nutritional economy of a household. She is the ultimate sustainable-energy vehicle.

The cow does not just provide protein for the other critters on the place. The effect of cow manure on the garden is magical. I go to very little trouble with composting, just starting a new pile occasionally and using up the old one. My garden soil is dark and friable and grows strong, healthy plants with a minimum of effort.

Can you put a price on all this? Maybe. I often read articles, books, and newsletters with suggestions on how to spend less on "lifestyle" so couples can get along on one income. Would home production of virtually all of your food make this possible for you? It will certainly keep you all radiantly healthy.

And another thing. Thrift is my middle name, but those suggestions for feeding the family on day-old bread and making bulk purchases of dry cereal I find depressing. Keeping a cow is more satisfying. One popular writer encourages frugality so that there can be savings in readiness for the children's orthodontia. I do not find this to be an incentive. If your children are young when you get a cow and they grow up with fresh milk, their teeth will be straight, just as all teeth were meant to be. How much we personally have saved on dental or medical bills would be difficult to state because health insurance costs vary. My emphasis has always been on prevention, not cure. Consequently cure has seldom come into it, but when it does it has a better chance in an already strong constitution built on real food.

Anybody Can Do This

Building family health affordably in an urban or suburban setting by shopping at whole-foods stores and farmers' markets and exercising iron self-control when surrounded by arrays of shining products should in theory be possible. I don't happen to have seen it done without at least keeping chickens and a vegetable garden. It can certainly be done with goats as the ruminant of choice, although goats offer fewer value-added options and often are more work than a cow. The cow is the most bountiful provider.

If you build family health on the basis of your own soil and animals, you need not worry overly much about the twenty-first-century degenerative diseases. You and your family can have the long-lasting health and good looks you deserve.

Restrictive diets stunt the growth of children and take the satisfactions out of old age. Advocates of restricted diets foster the impression that if you will cut out enough protein and fat and take enough of the vitamin-of-the-week you'll live forever. So far this has not occurred. What we have achieved is sickly old people and flabby youth. This outcome has not come about from a diet of milk, cream, butter, eggs, and meat, consumption of which has been declining steadily since around 1880. What has increased is consumption of manufactured fats, sugar, and refined carbohydrates. The people who built the stone walls and huge barns of America's farms did not do it on white bread and margarine. Nor were they assisted by vitamin tablets and capsules. Those vital constituents were found in their food, and they remain in the food of people who produce their own.

I cannot discuss the cost and work of keeping a cow without also considering the true long-term investment in the health and appearance of my family. The cost of my labor cannot be counted in this domestic economy. Nothing else I might have done with my time could have matched the rewards I see.

Cows and grass are recession-proof and inflation-proof. In difficult times, the family with a cow is not poor.

APPENDIX

Dairy Cow Anatomy

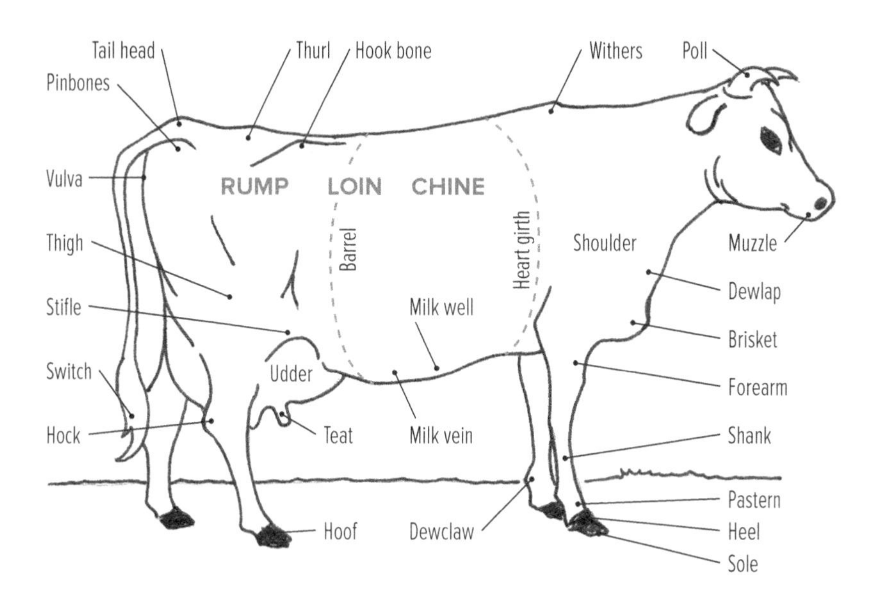

Sources

Cattle and Dairying Supplies

The three companies listed here are good sources for most cow care and dairying supplies, including weight tapes, anti-kicking devices, calf feeding bottles and hutches, some medications, fencing, and all types of milking and cleanup supplies.

Jeffers Livestock
800-533-3377
www.jefferspet.com

Lehman's
888-438-5346
www.lehmans.com

Premier
800-282-6631
www.Premier1Supplies.com

Cheese Making

New England Cheesemaking Supply Company
54B Whately Road
South Deerfield, MA 01373
413-397-2012
info@cheesemaking.com
www.cheesemaking.com
For quality cheesecloth and all cheese-making supplies

Ruminant Pregnancy Test

BioTracking
1150 Alturas Drive, Suite 105
Moscow, ID 83843
208-882-9736
www.biotracking.com
BioTracking is the developer of BioPRYN, a blood pregnancy test for cattle, bison, goats, sheep, deer, elk, and other ruminants.

A1/A2 Milk Test

Veterinary Genetics Laboratory
University of California at Davis
P.O. Box 1102
Davis, CA 95617-1102
530-752-2211
www.vgl.ucdavis.edu/services/A2Genotyping.php
The VGL is certified by the A2 Corporation Ltd. to offer a DNA test to identify animals that carry the A2 beta-casein variant.

CoPulsation Milking Systems

L. R. Gehm LLC
9502 NYS Rt. 79
Lisle, NY 13797
607-849-3880
www.copulsation.com

Animate Dry Cow Pellets

Prince Agri Products, Inc.
229 Radio Road
P.O. Box 1009
Quincy, IL 62306
800-677-4623
info@animate-dairy.com
http://animate-dairy.com/

Bibliography

Caldwell, Gianaclis. *Mastering Artisan Cheesemaking* (White River Junction, Vt.: Chelsea Green, 2012)

In-depth guidance with detailed illustrations takes you from beginner to professional cheesemaker. Includes milk chemistry.

Carroll, Ricki. *Home Cheese Making* (North Adams, Mass.: Storey Publishing, 2002)

The classic text covers all you need to know and where to order supplies.

Cheeke, Peter R. *Impacts of Livestock Production on Society, Diet/Health and the Environment* (Interstate Publishers, 1993)

An authoritative source of useful information about animal products and their role in health and disease. Cheeke clearly distinguishes between proven fact, reasonable conjecture, and his personal opinions. References are to published research in peer-reviewed journals. The style is sprightly and accessible.

Cobbett, William. *Cottage Economy* (White River Junction, Vt.: Chelsea Green, 2012)

Nineteenth-century farming advice that is largely relevant and always interesting.

Fairlie, Simon. *Meat: A Benign Extravagance* (White River Junction, Vt.: Chelsea Green, 2010)

A definitive study of the relationship between land productivity and human requirements. Fairlie's wry wit and many "aha" moments made this a page-turner for me. Meat wins.

Fallon, Sally, *Nourishing Traditions: The Cookbook That Challenges Politically Correct Nutrition and the Diet Dictocrats* (San Diego, Calif.: ProMotion Publishing, 1995)

Contains many references illuminating the commercial/political influence on reigning nutrition dogma. The recipes are interesting and use lots of dairy products. The sidebar facts and quotes are highly informative.

Gumpert, David E. *The Raw Milk Revolution* (White River Junction, Vt.: Chelsea Green, 2009)

Well-researched text by an investigative reporter whose research has been invaluable to the raw milk movement.

Hasheider, Philip. *The Family Cow Handbook* (Minneapolis, Minn.: Voyageur, 2011)

Detailed information on cow health and calving management; beautiful photography.

Hutjens, Mike, and Earl Aalseth. *Caring for Transition Cows* (Fort Atkinson, Wis.: W. D. Hoard & Sons, 2005)

Enhanced detail on metabolic disorders and feeding plans for those who want to know more.

Karlin, Mary. *Artisan Cheese Making at Home* (New York: Ten Speed Press, 2011)

Easy-to-follow instructions and inviting recipes for both beginners and advanced cheese makers.

Kimball, Kristin. *The Dirty Life* (New York: Scribner, 2010)

Modish New Yorker transitions into deeply committed farmer's wife. Her observations are astute and her style absorbing, and she is one heckuva good sport.

Logsdon, Gene. *All Flesh Is Grass* (White River Junction, Vt.: Chelsea Green, 2004)
The Contrary Farmer (White River Junction, Vt.: Chelsea Green, 1994)
Holy Shit (White River Junction, Vt.: Chelsea Green, 2010)
Small-Scale Grain Raising (White River Junction, Vt.: Chelsea Green, 2009)

Don't miss anything by Gene Logsdon.

MacGregor, Charles A. *Directory of Feeds and Feed Ingredients* (Fort Atkinson, Wis.: Hoard's Dairyman, 1994)

Contains many useful tables.

McCullough, Marshall E.. *Feeding Dairy Cows* (Fort Atkinson, Wis.: W. D. Hoard and Sons, 1986)

Further explanation of rumen function.

Proulx, E. Annie, and Lew Nichols. *Dairy Foods Cookbook* (Emmaus, Penn.: Rodale Press, 1982)

Marvelous information and recipes for preparing all kinds of dairy products. No sugar is used, only honey.

Salatin, Joel. *The Sheer Ecstasy of Being a Lunatic Farmer* (Swoope, Va.: Polyface Inc., 2010)

Joel Salatin leaps the tallest buildings as he demonstrates how to be a financially successful farmer within the rules of nature. All his books are inspiring.

Schmid, Ron. *The Untold Story of Milk* (Washington, D.C.: NewTrends Publishing, 2009)

Irreplaceable information on the history and science of raw milk.

2 Million Years of the Food Industry (Nestlé S.A., 1991)

Published by Nestlé to commemorate the Swiss company's 125th year. The book is a well-researched and superbly illustrated history, beginning in the Paleolithic period and reaching to the present, describing the foods of many lands. The authors (unnamed) present archaeological evidence establishing the keeping of animals for milk at least thirty thousand years ago, thus avoiding the common error of linking dairying to the rise of arable (grain) farming a mere ten thousand years ago.

Ussery, Harvey. *The Small-Scale Poultry Flock* (White River Junction, Vt.: Chelsea Green, 2011)

Everything you need to know about keeping poultry. Everybody needs chickens.

Weaver, Sue. *The Backyard Cow* (North Adams, Mass.: Storey Publishing, 2012)
Hands-on cowmanship and beautiful photographs of rare breeds of cows.

Williams, Roger. *Nutrition against Disease* (New York: Bantam, 1973)
Timeless and highly readable nutrition text.

Wilson, Duff. *Fateful Harvest: The True Story of a Small Town, a Global Industry, and a Toxic Secret* (New York: Harper Paperbacks, 2002)
Of this book Bill McKibben said, "This book is Erin Brockovich squared. Read it now."

Woodford, Keith, *Devil in the Milk* (White River Junction, Vt.: Chelsea Green, 2007)
Full explanation of our knowledge to date on A1/A2 milk, plus a look at the interface between science, politics, and corporate profit as it relates to A1/A2 milk.

Websites

www.familycow.proboards.com
Discussion forum: invaluable immediate help from experienced and caring cow owners.

www.keepingafamilycow.com
Author's website.

Index

NOTE: Page numbers in *italics* refer to figures and photos. Page numbers followed by *t* refer to tables.

About the Author

Sally McGuire

Joann S. Grohman has been farming and gardening organically since 1950. She and her family have six decades of experience raising poultry, goats, and both dairy and beef cattle in the United States and in England, where she and her husband, Merril, once owned and operated a 120-cow Jersey dairy in Sussex. In addition to *Keeping a Family Cow*, she is the author of *Born to Love: Instinct and Natural Mothering* (1976) and *Real Food* (1990). She lives and farms in western Maine.

About the Foreword Author

Jack Lazor has been farming since 1976 and is co-owner of Butterworks Farm in Westfield, Vermont, with his wife, Anne, and cofounder of the Northern Grain Grower's Association. Lazor grows grains both for human consumption and for feed for their herd of Jersey cows.